WILDLIFE LAW

WILDLIFE LAW

A Primer

Eric T. Freyfogle

Eric T. Freyfogle

Max L. Rowe Professor of Law
University of Illinois College of Law

Dale D. Goble

Margaret Wilson Schimke Distinguished Professor of Law
University of Idaho College of Law

Library of Congress Cataloging-in-Publication Data

Freyfogle, Eric T.
Wildlife law : a primer / Eric T. Freyfogle, Dale D. Goble.
p. cm.
Includes bibliographical references and index.
ISBN-13: 978-1-55963-975-0 (cloth : alk. paper)
ISBN-10: 1-55963-975-X (cloth : alk. paper)
ISBN-13: 978-1-55963-976-7 (pbk. : alk. paper)
ISBN-10: 1-55963-976-8 (pbk. : alk. paper) 1. Wildlife conservation—Law and legislation—United States. I. Goble, Dale. II. Title.
KF5640.F74 2008
346.7304′6954—dc22

2008022559

Printed on recycled, acid-free paper 

Manufactured in the United States of America

10 9 8 7 6 5 4 3 2 1

For Jane
and for Susan

Contents

Preface

Wildlife law is at once among the oldest and the newest fields of American jurisprudence. It is old because the first English-speaking immigrants brought to the continent a centuries-old body of law, stretching back to the Norman Conquest, which viewed wild animals as important and worth protecting. It is new because concerns for wildlife have broadened appreciably within the past generation, bringing corresponding, extensive changes in the rules governing animals and human interactions with them. Wildlife law still addresses long-familiar issues of hunting, fishing, and owning animals. Now it also reflects desires to protect biodiversity and land health, to respect animal welfare, and to confront exotic species and wildlife pests. Never has the subject been so engaging and vital.

This book provides an overview of wildlife law in the United States. We have attempted to make it both legally sound and accessible to readers with no legal training. Our target audience includes professionals and students in the wildlife and natural resources fields, as well as citizens concerned about conservation, wild animals, and the health and beauty of landscapes. Law students and lawyers, though, should also find the book useful. Despite wildlife law's long history, this is the first volume to offer something approaching a complete overview. Other books have reviewed federal or state statutes, examined stages of the subject's engaging history, and addressed specific topics such as the Endangered Species Act and wildlife refuge management.

Few works have exposed the subject's foundations, particularly the complex issue of wildlife on private land. None has brought the field together.

Our book, we believe, is timely given the rising importance of law in American society and the unavoidable need for people to learn about it. Professionals in most fields can hardly do their jobs without knowledge of the law's implications. Wildlife-related professionals, including the full sweep of land managers, fit within this group. Laws govern wildlife on federal and state lands and set rules for game ranches. They guide resource-management practices and oversee hunting easements. In one way or another, they address both over-abundant species in suburbs and our collective efforts to avoid extinctions. Increasingly, conservation practices are emphasizing the ways wild species promote ecological functioning; as they do, wildlife law is taking on new forms and gaining even greater importance.

This book should engage the interests of a wide range of Americans. The law is more than a collection of rules. It is a repository of our history and hard-earned wisdom. At the same time, the law offers a window into divisive cultural clashes and the practical challenges of living in nature. We write, to be sure, as biased observers. But in our view no legal subject offers more intriguing insights on people, land, and the sweep of history. Certainly few are so full of stories. We've done our best to give American wildlife its due.

The organization of the book is as follows. The initial chapter introduces readers to our legal system, to the types and forms of law, and to the various aims of wildlife law. It is intended to de-mystify legal culture and enable untrained readers to engage the field. It also sets the stage by probing the many functions wildlife law performs. The second chapter enters the field's substance by considering the foundational "state ownership" doctrine and the various public interests at stake. From there, attention turns to the basic rules—how a person becomes first owner of an animal; the many limits on capture; and the kinds of property rights owners obtain. Chapter four digs into a central challenge of all wildlife law: the uncertain status of publicly owned wildlife on private land. The tension here is evident: the public for many reasons wants wildlife protected while landowners often seek maximum freedom to control and use their lands as they see fit.

Over the centuries, the most economically valuable animals have been fishes and shellfish. Most early wildlife law dealt with them. We turn to the topic in chapter five by exploring the rich history of inland fisheries, including the confusing issue of public access to waterways. Rivers, of course, cross state boundaries. So do waterfowl and many other animal species. Chapter six incorporates this complexity by considering how power is allocated among levels of government and the specific constitutional limits on what particular governments can do. The legal rules considered here are essentially of the house-keeping type: they divide up the tasks and keep governments from working at cross purposes. With that framework in place we turn to the specific laws that various governments have crafted. Chapter seven surveys *state* laws governing hunting and fishing and establishing liability rules for harms caused by wild animals, in and out of captivity. Chapter eight takes up the extensive, contentious subject of hunting and fishing rights held by *Indian tribes,* on and off reservations. Finally, chapter nine explains four of the most important *federal* statutes, enacted by Congress to aid states in managing wildlife and to offer direct protection for species of special concern.

The final four chapters rise above conflicts over specific animals and specific lands and rivers to look at entire landscapes and species. Wildlife can survive only if populations remain healthy. That can happen only if wildlife have suitable habitat, often expansive. Chapter ten considers wildlife on the four systems of federal land that together comprise some 96% of all federal holdings: the National Forests, the U.S. Wildlife Refuges, the National Park system, and the public-domain lands administered by the Bureau of Land Management. Here, issues of hunting and fishing are eclipsed in importance by issues relating to land-planning at large spatial scales. Chapters eleven and twelve take up the Endangered Species Act, which stands at the front edge of legal efforts to revive imperiled populations and to give wildlife due consideration in land-use decisions.

The book ends, appropriately, by looking ahead. We have no crystal ball and cannot predict future lawmaking efforts. It is nonetheless possible to clarify key issues and frame our basic choices. New ideas are needed, for instance, to help conserve publicly owned animals that reside on private lands; at the moment, prevailing ideas on private land ownership provide little flexibility to protect the public's wild assets. Even more needed are improved methods of citizen governance that pay serious attention to nature's features and that invite the participation and support of rural landowners. In the end, wildlife will flourish only if people make it happen. They will do so only when and as they endorse the many benefits healthy wildlife populations can yield. The future of the law, then, depends upon prevailing culture, on ecological understandings, and on the wisdom and passion of advocates for the wild.

CHAPTER

1

The Basics

Wildlife law is the body of legal rules and processes that have to do with wild things—with the interactions people have with wild things, and with the interactions among people themselves that relate to wild things. It is as simple and as wide ranging as that. The term "wild things," of course, is a vague one, just as the term "wildlife" is vague, but lawmakers have ways of taking vague terms and giving them greater precision, even when the precision is arbitrary. For the most part, the law considered in this book deals with animals. It does so because human disputes over wild things have largely dealt with animals. The law pays far less attention to plants, and then usually just to domesticated and human-engineered ones. Among wild plants, only the rarest and weediest show up in the law. This divide reflects a fundamental difference in the law: plants are part of land (and thus belong to the landowner), while animals are not, a point we shall address. As for the many single-celled organisms that are neither plant nor animal, they also make few appearances in the law's annals, save as minor characters in disputes about diseases and, more recently, intellectual property rights. All life, of course, depends on these single-celled creatures. The law, though, gives them scant attention, and so shall we.

The place to begin any study of law is with the basics of law itself—where it comes from, why it exists, what forms it takes, and how it changes. So accustomed are we to the idea and presence of law that we rarely take time to think about it. The law that we're talking about, of course, is human-created law. Some person or group, somewhere at some time, created the law, and did so for particular reasons. Sometimes the reasons are unclear, and the law can seem mysterious. But all law exists for a purpose. When we understand the law's purposes, it becomes easier to understand the law itself. Much law deals with problems that arise among people. Other laws help people accomplish tasks that would otherwise not be possible. Then there are the various laws that allocate power or authority within society and that prescribe how that power can be exercised. Each type of law appears in the pages that follow.

LEVELS OF GOVERNMENT

One characteristic of law is that it is crafted by a variety of governments at multiple levels. The basic unit of governmental power in the United States is the *state.* The states, in essence, replaced the British king and parliament at the time of the American Revolution. States possess that all-purpose governing power commonly (though not entirely accurately) known as the "police power," the power to enact laws promoting the public health, safety, and welfare. Although limited by constitutions in what they may do and how they may do it, states are our governments of fullest sovereign authority. The *federal government,* in contrast, is a government of enumerated powers. That is, it possesses only those powers granted by the United States Constitution. Finally, there are the *local governments,* which derive their powers from a state. They typically possess only those powers expressly delegated to them, although the powers of "home-rule" jurisdictions can be broad. An action by local government that isn't authorized by state law is, for that reason alone, legally invalid.

Before considering other governmental bodies that play lawmaking roles, it is worth noting that all levels of government participate in developing wildlife law. Most wildlife law emanates from the state. Although local governments also enact laws dealing with wildlife ("ordinances," they are usually termed), their power to do so is typically limited. Local laws commonly address matters of public health, safety, and land use, not the management and conservation of wildlife populations. Federal law is also limited when it comes to wildlife—due to the conscious choice of federal lawmakers—although the federal role has grown. Federal law typically deals with wildlife issues that transcend state boundaries—either because the animals are migratory, or because violators of laws themselves cross state boundaries, or because problems arise that states alone either cannot or will not address. While less expansive, federal law preempts inconsistent state law and thus takes priority in case of a conflict.

For the same reason that wildlife law at the federal level has been on the rise, so, too, wildlife law has increased at the international level, by means of international treaties. Many of the problems addressed by international law deal either with migratory animals or with animals that inhabit the vast, unowned oceans. More recently, international law has been called upon to help enforce wildlife laws that individual nations adopt. Many violations of national laws are motivated by the profits people can make selling wildlife in international markets. Realistically, it is difficult for individual nations to enforce their internal laws unless global wildlife markets are tightly controlled. This task can be addressed effectively only by international agreement. Also fueling international wildlife law are widely held ethical and aesthetic concerns about the plights of wild species worldwide. Many people feel strongly that humankind should protect all species, preferably in their native habitats. People in one country can express great interest in the status of wildlife in other lands. This interest can also prompt nations to enter into agreements protecting particular species.

Among the sources of American law are the many special government bodies that administer statutes and ordinances. Commonly termed *administrative* or *regulatory agencies*, these bodies have varying degrees of legal independence. They also have varying powers to interpret, supplement, and implement the laws enacted by legislatures. Like local government bodies, administrative agencies possess only those powers expressly given them by a legislative body. Thus when questioning the validity of a regulation or action by a state fish and game commission or the U.S. Fish and Wildlife Service, the first place to look is the statute that created the agency and defined its powers. To use a typical example, the question often arises whether a state wildlife agency can designate a species such as the mourning dove a "game" species, thereby opening it up to hunting. As we shall see in chapter 7, legislatures in some states have empowered a state agency to make the decision. In other states, the legislature has retained the power, presumably because the issue is politically contentious. When a state legislature retains the power to make the decision, the state's fish and game agency cannot change the rules. Administrative agencies, in short, are obligated to stay within the legal powers, or jurisdiction, given to them.

Administrative agencies are also obligated to comply with the procedures set up to guide their activities. Typically, the most effective way to overturn an administrative action is to show that the agency failed to follow the proper procedures. Even if the agency's decision or ruling was within its jurisdiction—and, indeed, even if the ruling made good sense—it will be invalid if the agency skipped a required procedural step.

A final group of governments that needs mention is the *Indian tribes,* which hold sovereign powers over their territories and the wildlife within them. Many tribes also have rights to hunt and fish outside their reservations under treaties signed with the United States. Because Indian treaties have the same legal status as federal statutes, they take precedence over claims based on state laws, including the property rights of landowners. Many lands in which Indians retain hunting and fishing rights are now the property of non-Indians. Whether or not these landowners realize it, tribal members can sometimes lawfully enter their lands to hunt and fish without their permission.

FORMS OF LAW: THEIR GAPS AND OVERLAPS

Having surveyed the types of governments that create law, it is useful next to consider the various forms that law can take. The most binding and inflexible form of law is a *constitution,* whether federal or state. Constitutions are approved directly by citizens, and only citizens can change them. Like it or not, legislatures are obligated to follow constitutional mandates.

Next in priority come *statutes,* which are formal, written laws enacted by a body that possesses legislative power. Congress, of course, holds the legislative power at the federal level. State legislatures or general assemblies (the names vary) hold this power at the state level. At local levels, legislative bodies promulgate *ordinances.* Many names are used to describe these "municipal corporations," and, as we noted, they can possess various legal powers. City councils and county boards typically have wide authority to promulgate ordinances to promote the public welfare. More limited legislative power is commonly vested in park districts, forest preserve districts, coastal or tidal protection entities, public health districts, highway districts, and countless similar legal entities. In each instance, an ordinance is valid only if the entity promulgating it held the legal power to do so and followed proper procedures.

The countless *regulations* and *administrative rules* promulgated by administrative agencies make up yet another form of law. Administrative regulations are usually more detailed than statutes. Indeed, many administrative bodies are created precisely because detailed rules are required and legislative bodies lack the time and expertise to write them.

The form of law that most people find hardest to understand is the *common law.* This is a type of state law, but, unlike statutes or constitutions, it emerges directly from the accumulated decisions of state courts. This law originally was "common" in the sense that it applied throughout the realm of England, not just in a particular locale. Common law was the law applied by the courts of the kings and queens, as opposed to law that was peculiar to a given shire or privately owned manor. Over the centuries, England's royal courts decided a large number of disputes, and their decisions gained precedential value. As lawyers and judges studied these many judicial decisions, they pieced them into rules of law. These rules were viewed as guides for future conduct. For centuries, courts were reluctant to admit they were gradually writing new law as they handed down decisions. They liked to say instead that they were merely "finding" the law, or that they were clarifying law that already existed. But by the nineteenth century these subterfuges were largely discarded. Judges became more open about what they were doing and why. Increasingly, the common law was called "judge-made" law, to distinguish it from other forms of law.

The common law retains considerable importance, even though legislatures can freely change it. Because state statutes take priority, the common law remains valid only so long as the legislature does not alter it. Yet, in the wildlife arena, as in much of property law generally, many common law rules continue to govern disputes with only modest statutory alterations. The law of trespass, for instance, remains grounded in the common law. So do the rules governing liability for the harms caused by, or to, wild animals. The rule of capture, which specifies the actions that a person must take to obtain a wild animal, is also a common law rule, although now modified by fish and game laws. The rules governing hunting easements, to cite a further example, are also based on common law principles.

When legislatures do decide to amend the common law, they often change it in small ways. They may replace the common law, for instance, only as it applies to a specific factual setting. When this happens, the general common law rule remains in effect, as modified by the statute. To use an illustration: a legislature might enact a statute that allows hunters to enter posted private land to retrieve wounded game, while leaving in effect the common law rule of trespass that bars all entries onto private land without landowner consent.

As one can see from this discussion, it can be difficult to figure out what law governs a particular legal problem. Wildlife law is no exception. Even to resolve a seemingly easy question, a person may need to consult a variety of legal sources. We can illustrate this complexity by considering who has the legal right to fish in a stream, and when. State fishing regulations obviously need to be looked at, to see what fish can be caught, when, and how. To decide who can enter a particular water body to fish, we need to ask about the ownership of the land beneath the water. If the underlying land is public, then some government agency likely regulates its use. If the land is private, we would need to inquire about public rights of access, which are governed by a complex array of state and federal laws. Very briefly, on this last issue of waterway access: Members of the public can enter a water body if access is authorized by *either* state or federal law. We therefore need to consult both bodies of law. Any waterway open under federal law must remain open to the public; state law cannot close it. On the other hand, states are free to open waterways that are not open under federal law. The latter possibility is not viewed as a conflict between federal and state law since a state is not attempting to close a waterway that federal law keeps public. Often, as we will see, access issues turn on whether the water body is "navigable," a complex issue in itself.

This topic of stream access illustrates one of the most perplexing types of legal conflict that can arise. What law governs when different levels of government each provide a relevant legal rule and the rules differ? The usual answer, again, is that federal law overrides state law, just as state law, in turn, overrides local law. Or to generalize: law from the higher level of government is usually supreme in that it preempts the lower-level law. But this general arrangement is subject to exceptions. The chief exception is that Congress can authorize states to enact laws that would otherwise conflict with federal statutes. This exception comes into play in the instance of wildlife laws on federally owned lands. If it chose, the federal government could produce its own laws to govern hunting, fishing, and wildlife conservation on federal lands. These laws would displace state law. For various reasons, the federal government has not done so. Nor has it otherwise restricted the power of states to apply their wildlife laws on federal lands, except in national parks and other limited settings. As a result, people hunting on federal land are subject to state regulation.

One practical consequence of this division of lawmaking authority is that we end up with wide variations in applicable laws. Laws vary among the fifty states and among local jurisdictions within states. Even casual observers know that states differ in the species they consider game and the specific limits they place on capture. But the differences among states can be far starker. Consider, as an example, the zoos that governments operate. Zoo animals occasionally injure visitors. What facts must a visitor prove to obtain monetary recovery for an animal-inflicted injury? The answer varies widely. Some states view the harboring of dangerous animals as an "ultra-hazardous activity." In these states, the zoo is almost automatically liable when a dangerous animal injures a visitor, regardless how careful it's been. At the other end of the legal spectrum are states that grant zoos sovereign immunity from all liability, even for grossly negligent conduct. In such a state, an injured person would have no chance of recovery, no matter how careless the zoo's behavior. In between are states that allow injured visitors to recover upon showing that the zoo has been negligent in some respect. The variation among states could hardly be larger.

These differences pose a problem for any attempt to survey a field of law. With fifty states, we cannot note all the differences. Sometimes states collectively follow a single legal rule, or they may congregate around two alternate legal approaches. These instances are easy enough to summarize. On other issues, however, states embrace a wide array of approaches, leaving the law especially difficult to summarize. Making matters more challenging is that law is often vaguer than people realize. On many legal issues there is no clear answer. This is particularly true when a legal issue is governed not by a specific statute or regulation, but instead by the common law. Although the recorded decisions of courts now number in the tens or hundreds of thousands, new factual situations still arise daily. In such cases, it simply isn't possible to predict how a court would resolve the dispute—and, in the end, it is the ability to make such a prediction that enables us to state confidently the "law."

When it isn't possible to predict a dispute's outcome, the law is said to be vague. Adding to this inevitable vagueness is the fact that state courts are empowered to change the common law so that it comports with today's circumstances and values. We need to be careful, then, in claiming confidently to know the law, especially when the leading judicial decisions are decades old. Although a court will always give weight to earlier precedents, no matter how old, the older the precedent, the less weight it is likely to receive. Thus to know the law, one needs to anticipate legal change. It is no surprise, really, when a court today replaces a judicial ruling from a hundred years ago with a more up-to-date legal understanding.

One moderating force in all of this is the tendency of courts to pay attention to legal rulings from other states, to see what judges elsewhere think. A court in Georgia might look for guidance to relevant decisions from Texas or Pennsylvania, particularly if Georgia has no judicial rulings on a specific issue. The common law is not common throughout the United States, but courts sometimes give weight to rulings from other jurisdictions. When a specific legal issue has arisen recently in several states and the various courts have decided the issue uniformly, there is a good chance other courts will reach the same conclusion when confronting the issue.

AIMS OF WILDLIFE LAW

With these basic points in hand, we can take the next logical step, to consider the aims or functions that wildlife law performs. In general terms, legal rules serve to guide human conduct, either by resolving disputes or by letting people know where they stand legally so they can plan their affairs accordingly. What, then, does wildlife law do? What policies give it shape, and what aims are typically set for it? The more clearly we can answer these questions, the better we're likely to understand the various components of wildlife law and how they work. Aims also can give us a feel for the legal terrain. They help us anticipate the kinds of laws that might exist, while keeping us on the lookout for rules that might otherwise seem inexplicable.

Wildlife Conservation

For centuries, one major aim of wildlife law has been to conserve animals that people find valuable. The forms these laws have taken are as varied as the threats that wildlife faces. Early conservation laws sought to stem the overharvesting of animals by establishing closed seasons, bag limits, and gear restrictions. In time, as habitat losses became greater threats, lawmakers began creating wildlife refuges. More recent are the laws that protect wildlife against contamination, whether by toxic compounds or by ordinary organic wastes.

These laws are designed to protect the public's interest in wildlife. The reasons for conserving wildlife and, indeed, the meanings of "conservation," are diverse. Many wild species have gained protection because of their utilitarian value to humans. For example, game animals were protected so people could hunt and fish them. More recently, wild animals have earned protection because of the many other benefits they provide, whether ecological, aesthetic, or moral.

The various aims of conservation are important because in practice they shape the forms and levels of protection that wild species receive. If the wolf, for instance, is protected mostly for ethical reasons—because of a felt duty to protect the species for future generations to enjoy—then we might be content if we've adequately conserved it in a few geographic locations, even in zoos. If instead our interest is chiefly in the wolf as a symbol of wildness that enhances the experiences of recreational visitors, then our conservation goal might be to conserve the wolf in and around wild recreational lands. If we conserve wolves because of their roles in sustaining healthy ecosystems, then we are likely to take a much different protective approach. We might strive to restore the species to the ecosystems where it once existed or where it is now needed to keep prey species in check. Conservation goals, in short, are influential because conservation decisions are built upon them.

Much can be said on this subject, of course, and the goals of wildlife conservation are often quite contested. Sometimes it is the goal itself or its relative importance that is disputed. Sometimes the conflict comes when people who agree upon a goal—maintaining healthy fish populations, for example—disagree about the details or its application. When is a fish population healthy? How much harvesting can take place without impairing that health? These issues are often complicated by scientific uncertainty. To continue the fishing illustration, the amount of harvest that a healthy fish population can sustain is dependent upon a large number of factors whose interactions are often poorly understood. Even when everyone agrees on the conservation ideal and the desired biological outcome, it can prove difficult to know how to write the laws or how to apply them in specific situations.

Finally, some skepticism about the actual goals of actions described as "conservation" is often healthy. Most laws regulating harvests of wildlife are implemented in the name of conservation. Laws that limit harvesting methods, however, can in operation allocate wild animals to some people over other people—to sport fishers over commercial fishers, for example. Laws that regulate harvests often determine who or what type of person will get to harvest. The salmon fisheries on the Columbia River offer numerous examples of this effect. In the early twentieth century, Oregon banned fish wheels (a contraption that scooped fish out of the water) following an intense political struggle on the grounds that the wheels were "too efficient" and thus contrary to the state's conservation objectives. The ban, however, was directly allocational because it meant more fish for fishers who harvested in other ways. The relationship between wildlife "conservation" and "allocation" is often fuzzy—often intentionally so, as different users seek an advantage by characterizing their opponents' harvest methods as wasteful, unethical, or worse.

Allocation of Wildlife

As this discussion suggests, a second common objective of wildlife law is to determine who can acquire wildlife. On this issue, we might compare wildlife to other natural resources, because the same policy issues crop up, over and over. A chief aim of the law in the case of all natural resources is to make the resource available for human use. But how might this be done? What choices do lawmakers have when prescribing the ways people can gain private rights in a resource?

One very familiar method of resource allocation is sale to the highest bidder. Other methods include lotteries and first-in-time "capture" systems that give a resource to the first person to take it. These allocation methods, however, hardly begin to exhaust all the allocation possibilities. Resources can be allocated to those people deemed the most deserving, based on one factor or another. Alternatively, they can be allocated to foster public aims that are unrelated to wildlife. To use historic examples, land can be given as a bounty for military veterans or to induce railroads to construct new rail lines. Political processes and the perceived values of various harvesting methods are other factors that affect the allocation method that is ultimately chosen. In many settings, to use another common illustration, sport hunters or fishers have priority over commercial ones. Native peoples and subsistence families sometimes receive separate allocations. A licensing scheme for moose hunting might prefer a person who has never hunted moose over hunters who have already had a chance.

Licenses are commonly used by states to allocate rights to harvest wildlife. In many instances, licenses are limited in number to avoid overharvesting. When the number of licenses is limited, lawmakers must find a way to allocate the licenses. They can be handed out based on priority of application. They can be allocated by lottery among people who apply by a certain date. Often the residence of the harvester carries weight, although states do face constitutional limits on their ability to reserve wild animals for state residents. Finally, a particularly influential allocation method is one that operates indirectly: the law of property ownership that empowers landowners to control physical access to game-inhabited lands. Landowners must comply with wildlife laws, just like other people; there is no "right to hunt" on private land. Nonetheless, state trespass laws that allow landowners to control access in effect reserve wildlife on private land for the landowner. No one else can capture the animals so long as they remain there.

We can shed light on this point about landowner control by considering an old adage. It was formerly said that the United States embraced a far different approach to wildlife than did its legal parent, England. In England, wildlife belonged to the landowner. A person in England had to own land to hunt. Not so, it was said, in the United States, where wildlife belonged to the people and hunting was open to all.

This comparative summary did contain truth, particularly at the time of the Revolution. But it was not legally correct as presented. The English rule was that landowners had the legal power to exclude hunters. In addition, game killed by a trespasser went to the landowner, rather than to the hunter. In the United States, in contrast, the public had far greater rights to hunt, even on land owned by someone else. The custom in early America was that rural landowners could not exclude public hunters unless their lands were enclosed or tilled. So important was this public right to hunt on unenclosed lands, regardless of land ownership, that some states protected it in their state constitutions. Over the course of the nineteenth century, however, American law gradually changed. Step-by-step, in state after state, rural landowners gained the right to exclude public hunters. In Britain, landowners did not, in fact, legally own the wildlife on their lands; they simply had the right to exclude trespassers and seize any game that they captured. Today, wildlife law in the United States embraces essentially the same legal rule. Like so many children, we have grown up to look like our parent.

The point, of course, is that trespass laws, like many other laws, have indirect effects. Trespass laws do more than protect private land against unwanted invasions. They indirectly decide who gets to harvest publicly owned wildlife. The more rigorously we protect private lands, the more wildlife we effectively give to landowners as opposed to other citizens. In fact, we have become more protective of landowner rights. In most if not all states, trespass is a crime, so the power of the state can be used to punish a trespasser. In Britain, on the other hand, trespass is a civil matter, and a trespasser is liable only if he causes actual harm to the land or its contents.

Defining Private Rights

Closely related to the allocation of wildlife is another of wildlife law's functions: defining the rights a person obtains in an animal once it has been captured. What property rights attach to a captured wild animal, whether dead or alive, and how long do the rights last? Are they the same rights that attach to other forms of private property, or is wildlife different?

The law's answers to these questions are taken up in chapter 3. What is important to note at this point is that private property rights in wildlife are, in fact, quite different from other private property rights. Property rights in wildlife are more limited, even in the case of animals that are lawfully killed, reflecting the public's legitimate interests in all wild creatures. For example, an owner's property rights end if a captured animal escapes. In addition, people who lawfully kill wild animals often have no right to sell their property. They may even be limited in how long they can retain the animals or meat from the animals. Private property rights in wildlife are dependent upon law for their existence and scope, just as in the case with all other property. That is, owners possess only those property rights that the law prescribes, and states can redefine those rights, even after wildlife is caught. The bottom line is this: private rights in wildlife exist only to the extent the law authorizes the particular rights.

Ecological Considerations

Except in zoos and other captive situations, wild animals do not live in isolation. For centuries, however, the law treated wild animals as discrete pieces of the natural world. Some species were valuable, some harmful, and the vast majority fit into neither category. Today we are more likely to recognize that wild animals are embedded in and help sustain larger communities of life. When wildlife is viewed this way, the aims of wildlife law shift. It can be used to promote the healthy functioning of entire natural communities, rather than simply to increase or decrease populations of particular species of independent importance. Wildlife can help sustain a larger ecological community that is highly valued. Wildlife conservation, as a result, can be a stand-in for a full array of other conservation values and goals. Efforts to restore wolves in the northern Rockies, for example, have led to the reestablishment of aspen stands that had been overbrowsed by elk, once the elk learned that wolves favor cover.

According to many observers, ecosystems that support healthy populations of wildlife also provide an array of services to humans. Healthy wetlands, for example, slow the flow of water and reduce flooding. Wild bees and birds provide pollination services. Landscapes that support wildlife also provide recreational opportunities.

Another effect of viewing wildlife in ecological terms is that the economics of wildlife protection can shift considerably. When salmon are viewed as a harvestable commodity, the value of a protection effort can be measured by its effects on rates of harvest. When salmon are viewed instead as integral to healthy river systems, cans of salmon are no longer accurate as measures of benefits or conservation success. Indeed, steps to conserve wildlife can make good economic sense even if they *decrease* harvesting rates.

Pests

A longstanding aim of wildlife law has been to deal with animal populations viewed as pests. This aim has broadened in recent decades to include the problem of harmful exotic or invasive species. Early wildlife laws featured bounties for hunters who killed unwanted predators. Other early measures included bans on importing wild creatures into particular areas. Today, the suite of pest species has expanded well beyond predators to take into account the many other species that also cause problems. Pest control efforts now address species that disrupt ecological processes or that harm native inhabitants. This more ecological approach is new, but the idea behind it—that wildlife law is a tool for dealing with unwanted species—has a long and colorful history.

Ethical and Aesthetic Considerations

For generations, wildlife law has implemented society's ideas about beauty and the morally correct ways for people to interact with nature. Early thinking on this subject centered on wild species that appeared to possess particular values or that otherwise merited protection for reasons unrelated to any utility they had for humans. An early example took the form of special protection for the "royal species" of medieval England, including the majestic and conjugally faithful swan—protective rules that were skillfully parodied by novelist Herman Melville in *Moby-Dick.* Again, the child grows up to resemble the parent: in 1940, Congress enacted a statute protecting the bald eagle as "a symbol of American ideals of freedom."[1]

Historically, many wildlife laws have been designed to promote the ethics of sporting by banning harvesting methods that people consider unethical or unsportsmanlike. Regulations can prescribe, in effect, that wild animals have a fair chance of escape or that hunters avoid methods that make capture too easy. Laws can also seek to reduce the pain and suffering that animals endure before death. They can reduce the chance that an animal will be wounded but not killed.

Wildlife law today gives far more attention to the ethical and aesthetic implications of our dealings with animals. Efforts to protect rare species have been particularly visible and contentious. We protect endangered species mostly out of a belief that protection is morally important and because rare species can yield aesthetic enjoyment, whether or not they play critical ecological roles. For some people, species have inherent moral value that we ought to respect. For others, the protection of rare species is viewed as a duty owed to future generations. Yet another approach centers on concepts of virtue, good living, and national identity. To protect rare species is a form of humility; it is an exercise of restraint, whether by an individual or by a people. Merging with this point is the widely held belief that species protection is a religious duty, a duty to protect all God's creatures.

The serious literature on the moral status of wildlife is vast, growing, and, unfortunately, increasingly jargon laden. Species protection is a recurring topic of moral reflection, as is the moral status of entire biological communities. Drawing just as much attention has been the moral status of *individual* animals, rather than species or other collectives. The thinking on this subject, loosely termed "animal welfare," typically focuses on the pain and suffering animals endure at human hands. The law has long paid attention to this concern, usually in cases involving working livestock and companion animals. Anticruelty statutes have barred the ill treatment of captive animals and such "sports" as cockfighting and bearbaiting. Related statutes set standards for the humane treatment of animals in captivity, particularly in zoos, circuses, and research settings. Others impose limits on methods of slaughtering animals for food.

Anticruelty laws are now being extended, step by small step, to protect animals in the wild against various types of human conduct. This is an extraordinary legal development. Laws in some states now prohibit the capture of wild animals using methods that are deemed inhumane because of the pain and suffering involved. A common statute, for example, bans the use of leghold animal traps. In Colorado, the issue was deemed important enough to add to the state's constitution, which prohibits taking wildlife "with any leghold trap, any instant kill body-gripping design trap, or by poison or snare."[2] Aside from such new provisions, which show obvious concern for wild animals, there is a noticeable trend to include wild animals within the protective scope of longstanding anticruelty laws. Anticruelty statutes often protect "any animal," without further specification. It has been left to courts to decide whether the statutes apply to animals in the wild.

An example of how anticruelty law is broadening can be seen in legal shifts that have occurred in New Mexico. In a 1958 decision, the Supreme Court of New Mexico interpreted the term "any animal" in the state's anticruelty statute as applying only to "brute animals" and "work animals"—that is, working livestock that a farmer or rancher might mistreat.[3] In 1999, the same court interpreted a nearly identical newer statute as applying to *all* domesticated animals and wild animals in captivity, not just working livestock.[4] In response to the 1999 ruling, the legislature revised the statute so that it applied even more clearly to all animals, domestic and wild, but it then excluded fishing, hunting, and related activities as authorized by the state's fish and game laws.[5]

The apparent outcome in New Mexico is that all wild animals are now protected by the state's anticruelty law. That law prohibits not just maliciously cruel behavior, but "negligently mistreating, injuring, killing without lawful justification or tormenting an animal."[6] In 2007, the statute was extended to cover the controversial sport of cockfighting.[7] Although hunting and fishing are expressly excluded from the statute's coverage, it is unclear whether the exclusion includes *unlawful* as well as lawful hunting and fishing. Does the statute cover hunting and killing undertaken out of season or by unlicensed actors? A 1995 Indiana ruling, illustrates the questions and potential significance of such a statutory ambiguity.[8] An Indiana hunter who killed two Canada geese out of season was convicted of violating hunting laws in the case of a goose that he killed cleanly with a rifle shot. The other goose, however, was only wounded, and the hunter killed it by slitting its throat. According to the Indiana court, the act of slitting the goose's throat violated both the game law (hunting out of season) *and* the state's anticruelty statute.

The clear trend in this area of law is for states to expand the coverage of anticruelty statutes to include more types of animals. Newer laws also tend to prohibit broader ranges of harmful activities. Not just malicious actions or ones entailing great pain, but simply killing an animal, tormenting it, or even abandoning it after capture are sometimes now punished.

Miscellaneous Objectives

To this list of general aims for wildlife law we can usefully add a few further points. First, as we discussed in the chapter's initial section, many wildlife-related laws perform what might be termed housekeeping functions, in that they set forth what levels of government, and which government agencies, have power to act with respect to wildlife. Wildlife law is not a static body of rules that rarely changes. To the contrary, rules often vary from year to year, particularly the details about hunting and fishing seasons, bag limits, and no-hunting areas. Laws are needed to explain who makes such decisions, what processes must be followed, and how a person can challenge the rulings.

Second, many wildlife rules do not merely implement what many people view as basic rules of right and wrong conduct. It is thus wise to exercise caution when interacting with wildlife to avoid surprises from unexpected legal rules. This problem is particularly common in wildlife law because of the unusual challenges wildlife agents face in their law enforcement work. Only rarely does an agent actually witness an unlawful taking of game. It is equally rare that a wildlife agent has hard proof of the actual knowledge or mental state of a person at the time an alleged violation took place. For example, did the hunter *know* that the animal was a moose, not a deer, or that bait was laid out on the ground?

Recognizing these practical problems, lawmakers typically define wildlife violations more broadly than would otherwise seem fair. To enforce bans on killing, for example, laws also ban the possession of dead animals or even animal parts. Similarly, to enforce bans on hunting at night or with spotlights, states sometimes ban merely driving at night with car lights on and a gun in the vehicle, without requiring agents to show an actual intent to hunt. For decades, wildlife law has banned the possession of game out of season, without regard for whether it was lawfully killed during the season. Perhaps even more surprising are the legal rules protecting songbirds, which can ban the mere possession of feathers picked from the ground. Such rules are needed, it is said, if wildlife agents are to have a reasonable chance of enforcing wildlife laws. Because the prohibitions are so broad in coverage, though, it is easy to violate a statute inadvertently. As a result, it is vital that prosecutors use good sense when deciding what violations to prosecute.

RELATED TOPICS NOT COVERED

Like all bodies of law, wildlife law is knitted into a number of other bodies of law. Before we explore wildlife law systematically in the chapters that follow, it is useful to glance at some of these other legal fields, which are not included here.

One is the law that deals with livestock, pets, and other domesticated animals. It, too, enjoys a long history. The law of domesticated animals is more commercially focused because such animals are regularly purchased and sold, bred to produce offspring, and used as security for loans. A rich area of law, especially engaging for its insights on human society, is the law dealing with livestock that has wandered loose, or "estrays." It is easy to predict the concerns that underlie this area of law: the owner's desire to have an animal returned; the finder's desire to recover for any harm caused by the wanderer and to recoup the costs of capturing and tending the animal; and the public's interest in avoiding nuisances. The law of estrays has taken widely varied forms in different times and places. Beneath its surface are basic questions about livestock economics and local culture, such as the distinction between open-range and closed-range livestock grazing systems.

The line between wild and domesticated animals is by no means clear. The law, though, often must draw this line because legal rules apply differently to the categories. Particular problems arise when a species lives both in captivity and in the wild. This situation is becoming more frequent as landowners operate game ranches and raise wild animals in captivity. The typical legal approach is for the term "wildlife" to apply broadly. It includes all species that live in the wild, including individual animals held in captivity. In other words, animals are usually considered "wild" based on their species, not on whether they are confined and well behaved. The reason for this definition is apparent and again has to do with the challenges of law enforcement. Enforcement would be vastly harder if wardens had to show that a particular dead deer or duck was in fact wild rather than nonwild.

Whether a particular animal is or is not wild, and hence whether a particular wildlife law applies to it, often depends upon the exact wording of a statute or regulation. When the legal issue comes up, lawyers and courts must examine the specific law carefully to see how its terms are defined. Courts, in turn, must also keep in mind the common law basics about animals. When a captive wild animal escapes to something like its natural habitat, the owner's rights come to an end. When a domesticated animal escapes, on the other hand, the owner's rights continue unless the owner has intentionally abandoned it.

A good illustration of how these legal rules can operate appears in a federal appellate court ruling from 1961, *Koop v. United States.*[9] That case involved a landowner convicted of hunting migratory ducks over an area baited with corn. The landowner contended that the federal statute did not apply because the ducks being hunted were domesticated rather than wild and thus were excluded from the statute. The court agreed that the statute excluded "tamed or domesticated ducks."[10] But the mallards that the landowner raised and released were no longer privately owned once permitted to fly away. The law thus applied to them, the court decided, from the moment of release. In addition, the domestically raised ducks were indistinguishable from other ducks. The landowner had no practical way to identify the particular ducks that he had raised. All the ducks, therefore, were protected by the federal statute. This outcome was a sensible one, the court observed, because of the practical difficulties that would otherwise arise in enforcing the law. A contrary ruling on the issue would handicap enforcement of the law "beyond measure."

A similar case was *Schultz v. Morgan Sash and Door Co.,* a 1959 Oklahoma decision in which the court was called upon to decide whether state game laws applied to deer raised in captivity.[11] The plaintiff owned a 160-acre tract, protected by a high fence, in which he bred deer, buffalo, peacocks, and ducks, all for "recreation, enjoyment and amusement." The deer were fed from a barn and played with by visitors. Desirous of reducing the deer population, which had increased to fifty-five or sixty animals (including forty bucks), the plaintiff advertised for customers to shoot the bucks for fifty dollars each. The court rejected an attempt by state wildlife officials to regulate the activity. The state statute expressly excluded domesticated animals and pets; in the court's view the deer were "domesticated pets." In contrast to this ruling is a 2006 decision by the Oregon Supreme Court in *State v. Couch.*[12] There, a landowner imported three species of nonnative deer—fallow deer from Europe (made famous, the court noted, by Robin Hood's alleged poaching), axis deer from India, and sika deer from Japan. Did state game laws apply to these animals, acquired as captives and imported as private property? The court said they did, at least as long as the animals were unconfined on the private land.

The line that needs drawing in such cases, between wild and domesticated (or captive), is one that runs through much of animal law. Unfortunately, a line drawn for one legal purpose will often differ from the line drawn for another. An animal may be wild under one statute but not wild under another. It may be wild in one state and not in another.

Closely related to the wild-versus-domestic issue is the case of the feral animal, one that has escaped from captivity and lives in the wild. Where does it fit in the law? Once again, various laws give varying answers. An illustrative dispute is a 1964 Tennessee case involving feral hogs.[13] Here the issue was whether the hogs were covered by game laws that applied only to wild animals. The court concluded that the hogs were covered, given their wild behavior and reproductive vigor. Hunters were therefore subject to rules on taking game out of season.

Wildlife law also butts up against a number of other bodies of law, several of which can benefit wildlife indirectly. Pollution control laws protect wildlife by promoting the ecological health of lands and waters. Laws regulating toxic substances, including the pesticide control laws enacted in the wake of Rachel Carson's 1962 book *Silent Spring,* work similarly. Water pollution regulations are starting to use wildlife as indicators of the desired water quality for lakes and rivers. Under these "biological water-quality standards," waters are deemed clean only when they can sustain populations of particular aquatic species. When water bodies cannot sustain the desired mix of wild species, then tougher pollution control requirements take effect.

A final body of law that is largely excluded from this survey includes statutes dealing with the financing of wildlife habitat acquisition and other wildlife conservation measures. This law is difficult to wade into because it is linked to the complex legal rules prescribing the fiscal powers of government to spend money and to shift funds from one purpose to another. Among the federal laws in this area is the Pittman-Robertson Act of 1937, which provides financial assistance to states for "the selection, restoration, rehabilitation, and improvement" of wildlife habitat.[14] Nearly identical is the Dingell-Johnson Act of 1950, which provides funds for fish habitat.[15] Broader in scope than these statutes and not limited to the acquisition of wildlife habitat is the Land and Water Conservation Fund Act of 1963.[16] Readers interested in these statutes can find information about them elsewhere.

CHAPTER

2

State Ownership and the Public Interest

Perhaps the most conspicuous element of wildlife law is the peculiar legal status of animals living in the wild. According to courts, animals are owned by the people collectively, with the state acting as managerial trustee. This is an odd if not unique legal status. It provides the logical place to begin examining the substance of wildlife law. We need to see how this status evolved historically, why the arrangement still exists, and what it means in practical terms. What is the public's interest in wildlife? What powers does the state have to protect wildlife? And can citizens force the state trustee to take action when wildlife is being harmed?

THE PUBLIC'S INTEREST IN WILDLIFE

Wild animals are valuable parts of the natural world. They are important as commodities that people harvest and use. They are important also, as noted in chapter 1, for the functions they perform in natural systems. And they can also be accorded moral value or otherwise figure into the ways we evaluate right and wrong behavior, particularly when we look ahead to future generations. Ultimately, as Paul Shepard, a human ecologist, has argued, animals made us human: they were Others in a world "in which otherness is essential to the discovery of the true self."[1] Many states proclaim the public's interest in wildlife directly in their state constitutions, often as part of an expressed concern about natural resources generally. Pennsylvania's constitution, for instance, provides that the state's public natural resources are "the common property of all the people, including generations yet to come," and are to be conserved and maintained "for the benefit of all the people."[2] Several constitutions implicitly include wildlife in provisions protecting the environment generally. These provisions are often phrased in terms of the individual rights of state citizens. The "inalienable" rights of Montana citizens, for instance, include "the right to a clean and healthful environment."[3] In Massachusetts, the rights of citizens extend to "the natural, scenic, historic, and aesthetic qualities" of the environment.[4]

In some state constitutions, particular reference is made to hunting and fishing as valued activities. These provisions imply a state duty to protect game populations in perpetuity. Minnesota's constitution, for instance, declares that "hunting and fishing and the taking of game and fish are a valued part" of the state's heritage "and shall be forever preserved for the people," subject to management "by law and regulation for the public good."[5] Similarly, the people of Virginia "have a right to hunt, fish, and harvest game, subject to such regulations and restrictions as the General Assembly may prescribe by general law."[6]

When we look to England, the original source of our legal system, we find that the public's many interests in wildlife have drawn attention from lawmakers literally for centuries. The legal rule in medieval England was that game species were owned by the Crown, not by landowners or other private citizens. Game animals could be hunted or harvested only with the Crown's permission. Best known among wild animals were the few "royal" species—whales, sturgeons, and swans—which were highly prized by kings and queens for cultural reasons. But the Crown's claim to ownership of wild animals extended beyond royal species. Indeed, it covered all types of game, which is to say, nearly all wild animals of economic value. The medieval king's powers relating to wildlife encompassed the right to designate particular landscapes as royal forests. In such designated areas—many of them occupied by people and some not even forested—the Crown's ownership interests were even more extensive. Violations of the Crown's regulations brought stiff penalties. Special forest courts existed to try offenders.

The king or queen of England, of course, could not personally capture all the wild game of the realm. Instead, the Crown granted hunting rights to favored individuals, as a way of rewarding them for services rendered or securing their future support. Sometimes landowners received exclusive rights to hunt on their own lands. That practice, though, was by no means universal. The king could just as legitimately give the right to hunt to someone other than the owner. Needless to say, landowners could become annoyed when that occurred. Hunters knocked over fences, destroyed crops, and otherwise created nuisances. But while landowners might object, there was little they could do about it. More than one landowner was penalized for hunting on his own land, thereby violating hunting rights held by someone else. Even when landowners didn't face this problem, they might be unable to hunt on their lands because they didn't satisfy qualification statutes that limited hunting to citizens of considerable wealth.

In medieval England, the Crown was both the ultimate owner of all land and the head of government. These two roles and sources of royal authority overlapped. Particularly in the generations after the Norman Conquest of 1066, lawyers made little effort to distinguish between the two types of authority. Over time, however, it became important to draw lines between these powers, deciding what royal power was *sovereign,* or governmental, and what power was instead based on the Crown's *proprietary* ownership.

Many of the Crown's powers proved easy enough to place into one category or another. The royal courts and military forces, for instance, were exercises of sovereign powers; the income the king got from renting lands, on the other hand, arose out of the king's proprietary rights. Some of the king's powers, though, resisted easy division because they partook of both categories. The king, being greedy, was inclined to put as many powers as possible into the proprietary category. The king's power over that category was nearly unlimited. Parliament was inclined to do the opposite. It argued that various royal powers were sovereign powers, subject to management by law rather than royal whim. It made a big difference into which category a particular royal power fell. Over time, Parliament's view and its legal powers slowly gained ground.

This English history remains important today because it helps explain the legal status of wildlife in the United States and the seemingly odd way we describe it. All valuable wild species, English courts eventually decided, were owned by the king, but they were owned in a sovereign rather than proprietary capacity. This meant, importantly, that the king was obligated to manage wildlife in the interests of the entire realm, rather than for his personal benefit. And Parliament played a growing role in enforcing that obligation.

As it gradually solidified, this legal arrangement governing wildlife carried important consequences. Most visibly, it called into question the king's power to grant exclusive property rights in particular game populations. Most objectionable here was the controversial practice of granting to one person an exclusive right to hunt on lands that somebody else owned. Just as problematic legally was the practice of giving a landowner exclusive control over game found on his own land. The landowner could hold legal control over *access* to his land; that was becoming a basic element of land ownership, the right to exclude. But the wild animals themselves were subject to the ownership rights of the sovereign, of the king in Parliament (as the two components of the sovereign came to be known). Since the animals were not the king's private property, though, the public in theory could insist on their management for the benefit of all, not just for landowners.

These principles also governed fishing and shellfish harvesting in England. As owner of all lands beneath tidal waters, the king owned the fish and shellfish in the waters. Again, however, this ownership was a sovereign power rather than a proprietary one, or so courts finally decided after prolonged uncertainty. The Crown, therefore, was supposed to manage these fisheries to benefit the realm generally. Fisheries in *tidal* waters would thus be open to public fishing. As for fish in *nontidal* waters, adjacent landowners owned the underlying land and thus retained the power to exclude outsiders from coming in to fish, even when a river was a public highway. Because of their power over access, landowners could effectively control the inland fisheries. As in the case of dry land, however, their control was subject to laws enacted by the king in Parliament protecting the sovereign's interests in the fish.

This new legal arrangement left a knotty legal issue that English courts struggled with for generations: what was the legal status of wildlife that was unlawfully killed by a trespasser on private land? England's various courts came to differing conclusions on this question. It would remain unresolved until the House of Lords finally intervened in the 1865 case of *Blades v. Higgs.*[7] There the Lords ruled that a game animal killed by a trespasser immediately became the property of the landowner where it was killed. The Lords didn't reach this conclusion because the animal when alive was owned by the landowner; it emphatically was not. Their reasoning was far different. The key point was that the law "abhorred" the idea of property that was unowned. Every item that could be owned should have an owner, the Lords stated, women, children, and servants included. The sovereign's ownership of a wild animal, however, lasted only so long as the animal remained alive.

Once killed, the animal was no longer wildlife, or so the Lords assumed. Someone, then, had to become the new owner of the carcass. But who? The trespasser shouldn't own it because the trespass made his capture and ownership unlawful. With the trespasser out of the picture, the only option was to give the carcass to the landowner. And thus, said the Lords, the landowner became owner.

By way of sidelight, it is interesting to note that the ruling in *Blade v. Higgs* drew decidedly mixed reactions in the United States, which had a much stronger tradition of free public hunting in the countryside. In the view of some American judges, the trespassing hunter should get to keep the animal. The trespasser might have to pay for any harm due to the physical trespass. That was usually only a trifle if the hunter merely left footprints. He might also be exposed to criminal charges if the trespass violated a criminal law and if a busy prosecutor happened to take the case seriously. But if the trespass was trivial, it was too harsh a penalty to make the hunter relinquish the valuable game. Trespassing hunters were typically poor and needed the food. A slap on the wrist was penalty enough.

THE STATE OWNERSHIP DOCTRINE

In the United States, of course, the Revolution toppled the king. Sovereign power passed to the several states for their separate exercise. In time, American lawmakers embraced a rather uniform interpretation of the legal status of American wildlife, along the lines of English precedents. The states owned wild animals, lawmakers announced, but in a special legal way. They owned them in trust for the people generally and with a duty to manage them for the benefit of the many rather than the few.

By the end of the nineteenth century, ownership trust language of this type regularly appeared in American judicial rulings. Many state courts, as they endorsed this legal position, took time to explain its English origins. Typical of many discussions was one by the Washington Supreme Court in 1914:

> Under the common law of England all property right in animals *feræ naturæ* [wild animals] was in the sovereign, for the use and benefit of the people. The killing, taking, and use of game was subject to absolute governmental control for the common good. This absolute power to

> control and regulate was vested in the colonial governments as a part of the common law. It passed with the title to game to the several states as an incident of their sovereignty, and was retained by the states for the use and benefit of the people of the states, subject only to any applicable provisions of the federal constitution.[8]

As for the precise nature of the state's ownership, several courts explained it as carefully as they could. A much-cited discussion was penned by the Supreme Court of Minnesota in 1894:

> We take it to be the correct doctrine in this country that the ownership of wild animals, so far as they are capable of ownership, is in the state, not as proprietor, but in its sovereign capacity, as the representative, and for the benefit, of all its people in common. The preservation of such animals as are adapted to consumption as food, or to any other useful purpose, is a matter of public interest; and it is within the police power of the state, as the representative of the people in their united sovereignty, to enact such laws as will best preserve such game, and secure its beneficial use in the future to the citizens, and to that end it may adopt any reasonable regulations, not only as to time and manner in which such game may be taken and killed, but also imposing limitations upon the right to property in such game after it has been reduced to possession.[9]

California's Supreme Court made the same point more succinctly in a ruling from the same year: "The wild game within a State belongs to the people in their collective sovereign capacity. It is not the subject of private ownership, except in so far as the people may elect to make it so."[10]

Language of this type, declaring and explaining the state's ownership, also appeared in many state statutes and can even be inferred from state constitutions. Louisiana's highest court interpreted that state's constitution as establishing "a public trust doctrine requiring the state to protect, conserve and replenish all natural resources, including the wildlife and fish of the state, for the benefit of its people."[11] The Alaska Supreme Court has similarly concluded that its constitution, which expressly reserves the state's wild resources to the people "for common use," implicitly imposes upon the state "a trust duty to manage the fish, wildlife and water resources of the state for the benefit of all people."[12] As for the many state statutes expressly proclaiming the ownership doctrine, they are typified by those of Alabama ("title and ownership to all wild birds and wild animals . . . are vested in the state");[13] Arizona ("wildlife, both resident and migratory, native or introduced . . . are property of the state");[14] Colorado ("all wildlife . . . is declared to be the property of this state");[15] and Wyoming ("all wildlife in Wyoming is the property of the state").[16]

The state ownership doctrine received a ringing endorsement by the United States Supreme Court in 1896, in the prominent case of *Geer v. Connecticut.*[17] In that ruling, the Court surveyed English wildlife law in detail, emphasizing the long, unbroken power of government to regulate wildlife on behalf of the people. That sovereign power existed as "a trust for the benefit of the people, and not as a prerogative for the advantage of the government as distinct from the people, or for the benefit of private individuals as distinguished from the public good." This power passed from the English government to the original states upon independence and to the new states upon their entry into the Union.

At issue in *Geer* was whether the state's power over wildlife was sufficient to give the state free rein in prescribing the precise property rights that a person could obtain in a wild animal. According to Connecticut law then in effect, game birds lawfully killed within the state could be sold in the state but not transported and sold elsewhere. The law had the effect, if not the design, of discriminating overtly against interstate commerce, an activity that Congress rather than the states was charged to regulate. Despite this overt discrimination, the United States Supreme Court upheld the law's constitutionality. As the Court saw things, Connecticut owned the game birds while they were in the wild. It thus had full power to decide who could take them and when. That power included the ability to specify the precise property rights that a hunter obtained in a bird upon capture. The Connecticut law dealing with game birds, the Court ruled, did merely that: it provided that a person who lawfully captured a bird obtained only the right to possess, use, and sell the bird within the state's borders.

It wasn't long before *Geer* proved to be a troubling precedent. States soon realized that they held vast powers under *Geer* to engage in overt forms of economic protectionism. If they wanted, they could reserve all wild animals for use by state residents. At the time, seafood (fish and shellfish) possessed the highest market values. Various states prohibited out-of-state fishers from entering state waters or in other ways reserved fish stocks for capture by state residents. Some states mandated that out-of-state fishers unload their fish at docks within the state or have the fish processed by plants within the state. When states did allow out-of-state fishers to use their waters, they sometimes imposed higher license fees or added other requirements that placed out-of-state fishers at a serious disadvantage. In short, the state ownership doctrine provided a cover for express interferences with interstate commerce.

In a series of decisions during the twentieth century, the United States Supreme Court expressed growing doubt over the wisdom of *Geer v. Connecticut.* The ruling was specifically called into question in 1920 in *Missouri v. Holland,* an important decision upholding the federal government's power to protect migratory birds, even though Missouri claimed to own all birds within its borders.[18] "To put the claim of the State upon title is to lean upon a slender reed," opined Justice Oliver Wendell Holmes.

Things came to a head in *Hughes v. Oklahoma* in 1979.[19] In that decision the Supreme Court expressly overruled *Geer,* holding that the federal power to regulate interstate commerce precluded states from banning interstate shipments of wildlife. Much like Connecticut decades earlier, Oklahoma had a law prohibiting the export of minnows procured from the wild. On its face, the statute expressly discriminated against interstate commerce in a way that was patently unconstitutional in the case of any other commodity. But wildlife was different, or so the Supreme Court said in *Geer.*

According to the Supreme Court in *Hughes,* the state's attempt to discriminate in this way was inconsistent with the federal commerce clause. The state's power to specify property rights did not empower it to define rights in a way that discriminated against interstate trade. Although the state still held vast power to "protect and conserve wild animal life within its borders," it had to achieve those results in ways consistent with the recognition of the United States as an integrated economic unit. If it chose, a state could entirely ban trade in a wildlife item. There was no doubt about that. But having decided to allow commerce, the state couldn't define private rights based on state boundaries.

As a ruling under the commerce clause, *Hughes* was unexceptional. It merely brought wildlife into alignment with the law dealing with other market commodities. What spawned confusion was the fact that the Court in *Hughes* expressly discredited the whole idea that states really owned the wildlife within their borders. The idea of state ownership, the Court asserted, was merely "expressive in legal shorthand of the importance to its people that a State have power to preserve and regulate the exploitation of an important resource." Read literally, the Supreme Court in *Hughes* seemed to go far beyond reversing the narrow commerce clause issue in *Geer* to discard the entire state ownership doctrine. A number of legal commentators interpreted the ruling in that way.

The Court, however, lacked the constitutional authority to overturn the state ownership doctrine, and it is clear today that *Hughes* had no such broad effect. The ruling has been consistently interpreted as limited to the particular federal issue in the dispute: whether states can overtly discriminate against interstate commerce.

Since *Hughes,* several states have reiterated the basic elements of state ownership. As the Wyoming Supreme Court explained in 1986, seven years after *Hughes,* "Wildlife within the borders of a state are owned by the state in its sovereign capacity for the common benefits of all its people. . . . We hasten to [add] that the enlightened concept of this ownership is one of a trustee with the power and duty to protect, preserve, and nurture wildlife."[20] In a 1996 ruling, the Alaska Supreme Court made clear that *Hughes* did not in any way diminish "the trust responsibility that accompanied the state ownership doctrine."[21] A ruling by the Supreme Court of Montana in 1992 stated that *Hughes* was a valid precedent only on the specific constitutional issues in it; it did not alter the fact that wild animals were public property under the state's criminal law.[22] No state has seen fit, as a result of *Hughes,* to alter statutory expressions of the ownership doctrine.

THE IMPLICATIONS OF STATE AS TRUSTEE

The main effect of the states' sovereign ownership of wildlife is that states have extensive powers to protect wild animals. That power has been exercised regularly since the early days of American settlement, when colonies banned the taking of game species that were already becoming scarce. According to courts today, this power is largely unlimited.

When the legal issue first arose, it was less certain whether the state's power included the ability to protect wildlife on *private* land, particularly when the wildlife was doing damage. An influential ruling in 1917 by New York's highest court, *Barrett v. State,* gave an affirmative answer.[23] Using language about government power that was by then familiar, the court noted:

> The general right of the government to protect wild animals is too well established to be now called into question. Their ownership is in the state in its sovereign capacity, for the benefit of all the people. Their preservation is a matter of public interest. . . . Everywhere and at all times governments have assumed the right to prescribe how and why they may be taken or killed.

Barrett involved a complaint by a landowner about beaver damage done on his land, including the loss of two hundred trees. The beaver, then on the brink of extirpation in New York, had become the object of a concerted state conservation plan, including reintroducing the animal into areas where humans had driven it to extinction. As part of this conservation effort, New York enacted a law (no longer in force) prohibiting all persons, including landowners, from molesting or disturbing beavers or their dams and lodges. The landowner sued the state to collect compensation for the damage the reintroduced beaver had done to his land. In a unanimous ruling, the Court rejected the claim:

> Whenever protection [of wildlife] is accorded, harm may be done to the individual. Deer or moose may browse on his crops; mink or skunks may kill his chickens; robins eat his berries. In certain cases the Legislature may be mistaken in its belief that more good than harm is occasioned. But this is clearly a matter which is confided to its discretion. It exercises a governmental function for the benefit of the public at large, and no one can complain of the incidental injuries that may result.

Although the beaver was a new object of protection in New York, the court could see parallels with longstanding efforts to protect wild birds: "The prohibition against disturbing dams or houses built on or adjoining water courses," the court noted, was "no greater or different exercise of power from that assumed by the Legislature when it prohibits the destruction of the nests and eggs of wild birds when the latter are found upon private property."

Having decided that the state acted properly in protecting the beaver, the New York court had only to consider the fact that the state had affirmatively introduced the beaver into the area. The beaver had not returned to the region on their own. Did that make a difference? The court began its answer by recognizing that introduction efforts sometimes turned out badly:

> The attempt to introduce life into a new environment does not always result happily. The rabbit in Australia, the mongoose in the West Indies, have become pests. The English sparrow is no longer welcome. Certain of our troublesome weeds are foreign flowers.
>
> Yet governments have made such experiments in the belief that the public good would be promoted. Sometimes they have been mistaken. Again, the attempt has succeeded. The English pheasant is a valuable addition to our stock of birds. But whether a success or failure, the attempt is well within governmental powers.

If the introduction of an exotic species was permitted, the court reasoned, it was all the more proper for a state to return a native species to its former habitat. It was true, the court noted, that owners of wild animals were typically responsible for the harm they did, and the state in an important sense owned the beaver. But in "liberating these beaver," the court asserted, the state was not acting as private owner. Instead, it was acting in its sovereign, governmental capacity. "As a trustee for the people and as their representative, it was doing what it thought best for the interests of the public at large." Under such circumstances, it was not appropriate to hold a state liable for wildlife damage to the same extent as a private owner.

The state's sovereign power over wildlife has included not just the right to protect wildlife but also the power to seek legal remedies when animals are unlawfully harmed. This issue has most often arisen in connection with fish kills, caused by the intentional or accidental dumping of toxic substances. Does the state's ownership interest in wildlife include the right to recover money damages for dead fish, or is the state limited to collecting fines for violating water pollution laws?

At first, courts were unsure how to resolve this question, but after a few tentative and varied rulings they soon embraced the idea that state ownership does include this power. When fish or other wildlife are unlawfully killed, the state can recover money damages for the harm to state property. Once that issue was resolved, attention turned to the follow-up question, how to measure damages. What was the value of a dead fish or a dead otter? Some courts have looked to the market values that animals have when captured and offered for sale. The problem with that approach is that most species have no direct market value. An alternative valuation method has looked to the costs of replacing the dead animals. This method, too, has problems in that many animals cannot be replaced, particularly with animals of similar age. Yet another approach, not yet tested, is to calculate the money that hunters or fishers might pay to capture particular animals lawfully. How much is spent by fishers on average to catch a smallmouth bass or by hunters to kill a blue-winged teal? Adding to the problem of valuation is that the loss of certain animals—fish, for instance—can trigger the loss of other animals dependent upon them. Once begun, ecological ripple effects can spread widely and quickly become untraceable. How can a state recover for this damage?

So far, courts have been unable to produce a uniform method for valuing wildlife. In the typical case, the government proposes its own valuation method and the court goes along with it. Most governments have kept their monetary requests low, often using market or replacement values and ignoring the losses of species without market value. Because governments have been so modest, courts have rarely considered more aggressive valuation methods, such as those that include the full ecological ripple effects and the difficulty of replenishing many species except by letting nature take its course.

In thinking about the proper level of damages for harming wildlife, it may help to turn to basic economic principles. Harm to wildlife can be understood as a type of external harm (or "externality") that an actor generates and then imposes on publicly owned property. Externalities of this type skew the market's valuation processes. Economists commonly propose that those who generate externalities be forced to "internalize" them in the sense of having either to physically remedy the harm caused or to pay the full economic value of that harm. An alternative, related approach is to look to the profits or cost savings that an actor has incurred by avoiding the precautions needed to protect wildlife. To discourage an actor from doing the same bad thing again, economists propose that actors disgorge their profits or savings. That way, they end up no better off as a result of their violations. At the least, the law should remove the economic gains of harming trust property.

Related to this issue—and also largely unresolved—is whether the state is obligated to use money that's recovered to replace or repair the wildlife that's been harmed. Is the money that a state recovers earmarked for use only in promoting wildlife? Many times, it makes more sense to use the money, not to raise animals in captivity and then release them, but instead to improve the natural wildlife habitat. Nature can often replace the wild animals in time. Monetary awards might be better spent in enhancing its ability to do so. There is little reason why a state should not use money this way.

More troubling is the possibility that a state might simply put money recovered into its general treasury, or use it to fund conservation programs generally, then freeing up other tax money for use elsewhere. Again, the law on the subject is virtually nonexistent. The general idea of a trust, though, would obligate a state to use any money it gets to enhance trust property, which is to say the state's wildlife.

LOOKING AHEAD: TAKING THE TRUST DOCTRINE SERIOUSLY

Lurking behind the idea of wildlife as trust property, and indeed behind the whole doctrine of state ownership, is the often-repeated idea that a state owns wildlife as "trustee" for the people generally. It is hard to know how seriously courts are when they use this language. But the phrasing is heard so often that we need to take it at face value.

In legal terms, a trust is a distinct legal entity that holds title to property and manages it. The trust's ownership of the property, though, is a highly qualified one. Trusts are given property not to manage for themselves, but for the benefit of designated beneficiaries. Banks often serve as trustees, managing property for the benefit of one or more individual beneficiaries. The trust, it is said, holds "bare" legal title, while the "equitable" ownership of the property resides directly in the beneficiaries.

What, then, are we to make of the idea that states hold wildlife in trust for the people? Taken literally, such language indicates that states perform the role of trustee, with the people as beneficiaries. States thus must abide by the typical high standards of fidelity and honesty that apply to all trustees. They must scrupulously protect the trust property and, in the event of loss or damage, seek relief from those who have harmed it. The people as beneficiaries could presumably seek relief against a state if it failed to fulfill these duties.

The problem with taking trust language literally is that there is no trust document that sets forth the precise terms of the trust. Trustees are held to high fiduciary standards. They must scrupulously fulfill the terms of the trust. They particularly must safeguard trust property and manage it according to safe investment guidelines. If trust property is harmed, then trustees typically must seek compensation for the harm, filing suit if necessary and returning to the trust all money received. If a trustee fails to perform a duty, the beneficiaries can sue for malpractice.

What, then, are the management guidelines that a state should follow in the case of wildlife? What powers does it possess over wildlife, and when has it failed to fulfill its duties? Without a trust document, there are no ready answers. If answers are forthcoming, they will need to come from courts since the wildlife trust is itself a common law creation. Having created the trust arrangement, drawing upon centuries-old principles of common law, it is up to state courts today to give substance to the legal arrangement.

What makes this all confusing is that state legislatures have the powers to enact new laws, which generally take precedence over the common law. Thus the state legislature would seem to have the power by enacting new laws to change the legal terms of the wildlife trust itself. Or to put the point differently, the state as trustee would seem to have the power to alter the wildlife trust if it doesn't like the terms as they exist. In the law of trusts, however, trustees would never have this power: the power to change a trust lies with the beneficiaries, not the trustee.

So far, courts have had little or no occasion to struggle with these issues. The duties states have and the limits they face in managing wildlife remain largely undecided. This uncertainty can strike nonlawyers as distinctly odd, perhaps even unacceptable. But the truth is, the law at any time always has gaps in its coverage, often large ones. On many legal issues there simply are no answers. And answers are unlikely to come unless and until courts are pressed hard to give them, in lawsuits that arrive for resolution.

Looking ahead, it is rather easy to identify some of the legal issues that might press courts to flesh out the wildlife trust. One issue, already on the horizon, is whether a state is *obligated* to seek monetary compensation from a private party that unlawfully destroys trust property (that is, wild animals). Must a state exercise its power to seek damages, or can it instead sit back and do nothing? Under the law of trusts, trustees typically are obligated to act unless the harm to trust property is so slight that it makes no economic sense to seek relief. If we apply trust law to this issue, the answer is easy. But courts no doubt will go slowly. They realize that governments have a substantial amount of work to perform. Officials need to control how they spend their resources. Still, it makes no sense to give states complete discretion. After all, a suit to collect monetary damages is likely to increase state revenues.

If a state does fail to act, should private citizens as trust beneficiaries have the power to step in and do so on its behalf? On this question, too, we have no legal answers. Courts are likely to view the possibility with suspicion. They are reluctant to allow private citizens to enforce state laws unless the legislature has authorized it. On the other hand, the state ownership doctrine is itself a common law creation. If a citizen suit remedy of this type makes good sense, then it is appropriate for courts themselves to say so, without waiting for the legislature to act.

One significant way that trust property is harmed is through the loss of suitable habitat. Wildlife cannot survive if it has no place to live. What, then, might the trustee's role be in terms of protecting habitat? On this issue we might usefully distinguish between habitat on publicly owned land and habitat on private lands. Should a state have some legal obligation to ensure that wild species have places to live? Could a private citizen, as trust beneficiary, challenge state land management actions that shortchange wildlife needs? Questions such as these are at the forefront of wildlife law.

Perhaps the toughest of all wildlife law issues has to do with the status of wild animals on private land. As the court's opinion in *Barrett v. State* explains, states have vast powers to protect such wildlife. But is there an obligation to do so? No sooner is this issue raised than we see problems with it, in terms of the practical difficulties and the dangers of tampering with the property rights of landowners. Still, the public has property rights as well, not just the landowners. If they are less important than private rights in land, they are nonetheless far from valueless. It would be going too far to require that landowners manage their lands chiefly for the benefit of resident wildlife. On the other hand, it might make sense for landowners to bear some burden to leave room for native animals. Perhaps landowners should shoulder some fair-share obligation to help sustain local wild populations. (This issue is taken up in chapter 13.)

The state's ownership of wildlife, and its trustee duties to citizens generally, provide the legal foundation for the entire field of wildlife law. We shall need to keep these elements in mind as we proceed, step-by-step, through the legal field. Should courts at some point take this trust arrangement more seriously, elevating the standards that states must fulfill, they could alter wildlife law and benefit wildlife in significant ways.

CHAPTER

3

Capturing and Owning

With the broad framework of our legal system, the aims of wildlife law, and the concepts of state ownership and public interest in place, we now turn to some of the elements of wildlife law that most directly apply to citizens as they interact with animals. Our inquiry starts by considering the main way in which people gain ownership rights in wild animals, through the well-known rule of capture. We'll explore what actions qualify as lawful capture and then turn to questions involving private land: what rights do landowners have to exclude outside hunters, and who owns animals killed by trespassers? We'll similarly look at what happens when a capture takes place in violation of game laws. Assuming a capture is lawful and that a hunter gains rights in an animal, what are those legal rights, how long do they last, and what danger is there that lawmakers might redefine property rights in wildlife after capture has occurred? The chapter concludes by exploring special capture-related issues that arise on contemporary game ranches.

PIERSON V. POST AND THE RULE OF CAPTURE

No wildlife decision is more celebrated in American law, or better known to generations of law students, than a judicial ruling that ended a seemingly trivial spat involving a single fox on an uninhabited beach. The decision, titled *Pierson v. Post,* was handed down in 1805 by the New York Supreme Court in Queens County, New York.[1] The legal issue was the rule of capture. In its ruling, the New York court explained in legal terms what a hunter had to do to become owner of a wild animal. When the underlying drama began, the fox was not privately owned. The basic story therefore was about how private property rights arose: a physical thing (the fox) was unowned; a person came along and took it; and private ownership began. It's easy to see why *Pierson v. Post* has been, for thousands of law students, one of the first appellate rulings they read.

In *Pierson,* a young man, Lodowick Post, was hunting with his hounds on "unpossessed" land in the company of other hunters. While Post was in hot pursuit of a fox, another young man, Jesse Pierson, interrupted the hunt, killed the fox, and made off with the carcass. What happened next is factually unclear. We can imagine, though, that guns were brandished and profane words were unleashed. Also lost to time is why the dispute rose to such a high level, with the parties spending small fortunes hiring lawyers and litigating the case (over a thousand pounds each—well over a hundred thousand dollars in current funds). One historian's guess is that the two families already had bad blood between them. One family (Pierson) was of English background, the other (Post) apparently of Dutch; ethnic differences may have played a role. The young men involved, it appears, really were not the ones pushing the argument. It was their fathers—Captains David Pierson and Nathan Post—who may have had feuds of their own to settle. According to a more recent study, Pierson's family had lived in the area a long time and held rights to use the town's common "waste" lands, while Post's family were newcomers and lacked any such rights.[2]

Whatever the hidden facts and motives, Lodowick Post filed suit with a justice of the peace demanding return of the fox carcass. (The case went before a justice of the peace because the dispute involved property of such little value.) Pierson refused to give up the dead fox. Post won before the single justice, but Pierson promptly exercised his right to have the case retried before the full Supreme Court of Queens County. It was a court that featured an extraordinary group of lawyers, including a future vice president, two future U.S. Supreme Court justices, and a scholar soon to become the nation's leading author of legal treatises.

The legal issue was a simple one: what must a hunter do to gain property rights in a wild animal? Post's claim was that his hot pursuit of the fox was enough. He had started the fox with hounds, was close to capturing it, and had both the intent and ability to complete the capture. Pierson responded that none of this was sufficient. To capture an animal, Pierson argued, a person had to take physical possession of it. Nothing short of physical seizure would suffice.

Writing for the court's majority, Justice Daniel Tompkins sided with Pierson. Although a hunter might gain ownership by inflicting a mortal wound on an animal, Tompkins conceded, it was otherwise essential that a hunter actually take the animal into physical possession. No lesser act would qualify as capture. This legal rule made sense as a policy matter, Tompkins claimed, because the law otherwise would engender too many disputes. When was pursuit "hot" enough, and when was a hunter close enough? A "hot-pursuit" rule would simply be too vague to apply, leading to too many lawsuits. Even worse, Tompkins asserted, a vague, hot-pursuit rule could lead to on-the-spot disputes involving breaches of the peace. A clear rule that was easy to apply in the real world would work much better.

Writing in dissent, Justice Henry Brockholst Livingston sided with Post and his fellow fox hunters. The key policy issue, as Livingston saw matters, was one of fairness among hunters. If the dispute had been submitted to the arbitration of sportsmen instead of a court, they would have known how to resolve it. Pierson's action as interloper was unethical, if not despicable, and a jury of sportsmen would have said so. Hot pursuit with the apparent ability to capture should be enough to gain ownership, Livingston opined. A hot-pursuit rule would also encourage people such as Post to hunt foxes. The more encouragement people had to hunt foxes, the more foxes they would kill and the fewer chickens farmers would lose.

When law students encounter *Pierson v. Post,* it is presented as a case involving the origins of property. It is about the way a person gains ownership of an unowned thing. To view the case from that perspective, of course, is to see it from the viewpoints of the Piersons and Posts of the world, who are out to acquire private property. An equally useful way to view the case is from the point of view of society and its lawmakers. The fox was unowned, society apparently was happy to get rid of it (rather than retain it in sovereign ownership), and the easiest way to get the fox into private hands was to offer it to the first taker. The first one to grab the fox got to keep it. The New York court did not consider whether some other resource allocation method might better serve the public interest. The disagreement among the justices, therefore, was narrow. First-in-time, all agreed, was the governing rule. But first-in-time to do what? Was hot pursuit enough, or was physical possession required?

Pierson is a particularly interesting ruling because, while the facts seem almost timeless, the court was unable to find a single earlier precedent, from America or England, in which a court awarded a wild animal to a party simply for being first. Surely such disputes happened, and surely the first one to seize an animal physically ended up with it. But it was apparently true that the bare rule of capture was rarely applied in court. Hunting was an elite activity in England, reserved for those who held hunting rights from the Crown. The foxes in England therefore likely went to the well-dressed hunters who had legal rights to hunt them. As for situations where no one held hunting rights, the animal presumably went to owner of the land where it was found, unless the landowner expressly gave someone else permission to hunt it. Since all land in England was owned by someone, these two legal rules took care of pretty much all disputes. These realities explain why the New York court had such trouble finding factually similar precedents to guide its ruling.

In America, hunting was more freely available. No doubt, disputes over animal captures did arise with regularity. But when we search through the reported decisions of courts, we rarely come across cases involving such facts. One reason for this gap in precedents is easy to surmise. Litigating cases is expensive, particularly cases that go up on appeal and result in judicial opinions. Rarely is a single animal worth enough monetarily to make litigation economically sensible. Capture disputes, then, are most likely resolved in some other fashion, out of court.

When the economics of litigation are taken into account, it is no surprise that the animal that has generated the most court rulings has been the whale. Nineteenth-century American courts handed down a number of decisions dealing with whale captures. As they did so, the courts explained in detail what a whaler had to do to gain property rights in them.

Nineteenth-century whaling cases are interesting to study today. What they show is that courts paid close attention both to the practical realities of whaling and to the customs of whaling communities. One prominent legal decision involved a dispute over a finback whale that had been harpooned and killed near Provincetown, Massachusetts. When a finback died, it tended to sink, only to rise a few days later, miles from where it was killed. The question thus arose: did the whaling party that killed the finback get to claim ownership even though the party never took physical possession, or did the whale belong to whoever seized it once it arose?

The case came before the federal court in Boston in 1881, which ruled in favor of the original whaler.[3] To support its ruling, the court cited the established custom of the local whaling industry, under which a person who found a finback whale was supposed to send word back to Provincetown so the whaling party that killed the animal could retrieve it. The finder was compensated for the work involved but had to give up the whale. The federal court bolstered its ruling by discussing a number of other whale cases in which courts had similarly assessed the practical difficulties whalers had adhering to the precise capture rule in *Pierson v. Post.* A whale that was well marked with a harpoon or waif pole, particularly when these markers were held by irons, remained the property of the party that killed it, even though the dead whale might be alone on the seas when a subsequent whaler found it. If the markers were obvious enough, a whale might remain owned even if a storm disconnected it from its anchor.

Aside from the whaling cases, few reported appellate decisions have involved disputes over capturing individual animals. In one ruling from 1914, the Wisconsin Supreme Court gave ownership of a wolf to the person who had wounded it and who was so close to consummating the physical capture that escape was highly improbable.[4] Private property arose, the court decided, when actual possession by one hunter was "practically inevitable." A 1902 ruling from the Ohio Supreme Court applied the rule of capture to fishing traps.[5] The fish that entered the trap, the court ruled, became the property of the trap owner, even though any particular fish could conceivably escape. Aside from these cases, we have little more than a few decisions involving swarms of bees. There the legal rule was largely the same. Bees must be reduced to possession or something very close to it; it is not enough that a person is first to spot the swarm and run after it. For example, in the 1823 case of *Ferguson v. Miller,* the New York court held that it was the first person to reduce the bees to possession that became their owner rather than the person who was the first to discover them.[6]

The rule that emerges from these common law cases can be stated as follows: an animal that is lawfully available for capture (an important "if," of course) belongs to the first person to reduce it to actual possession. Mortal wounding is apparently enough if the hunter continues the chase. Short of that, courts consider the characteristics of the particular species, the specific challenges of hunting it, and any customs that apply widely among those who pursue the particular type of animal. When these factors are considered, it could well be possible that acts short of actual capture suffice to create private rights, at least if capture is nearly certain and if a more absolute rule of physical possession would burden hunters excessively or unfairly.

PROTECTIONS DURING THE HUNT

In the past decade or two, people opposed to hunting and trapping have occasionally disrupted hunters in their attempts at capture. Often the activists are animal welfare advocates who dislike hunting and trapping, either always or when conducted in particular places or ways. State legislatures have sympathized with hunters' complaints and have been remarkably quick to enact statutes banning "harassment." Such statutes have been enacted even in states with no reported incidents of harassment and where public sentiment is decidedly mixed about the ethics of hunting.

State statutes have taken various forms so as to avoid the possibility that a court might find that a statute has wrongly violated the free speech rights of the activists. A few courts, troubled by the constitutional implications, have struck down provisions of particular harassment statutes. For the most part, though, these statutes have withstood constitutional scrutiny, particularly when they focus on harassing actions rather than on speech intended to dissuade people from hunting, and when the conduct prohibited is directly related to the wildlife—scaring animals away—rather than aimed at the hunters.

In a typical case from 1998, the Illinois Supreme Court struck down as a violation of free speech rights a provision of the Illinois Hunter Interference Protection Act.[7] The statute penalized any individual who "disturbs another person who is engaged in the lawful taking of a wild animal . . . with intent to dissuade or otherwise prevent the taking." The constitutional defect of the statute, the court announced, was that it applied to speech based on the content of the speech. The statute applied only when the speaker had an "intent to dissuade"—that is, to convince the hunter to cease hunting. The court showed no concern about another provision of the same statute that banned physical interferences with hunting. The court's worry was solely about speech or other efforts to communicate with hunters to get them to change their minds. In a 2007 ruling, an Indiana appellate court had no trouble upholding the constitutionality of the brief Indiana Hunter Harassment Act, which applied to a person who "knowingly and intentionally" disturbed a game animal or otherwise affected an animal's behavior so as to hinder its taking.[8] A Connecticut court in 2002 upheld a similar statute, which provided that "no person shall obstruct or interfere with the lawful taking of wildlife by another person at the location where the activity is taking place with intent to prevent such taking."[9]

Even when the statute is constitutional, however, it still must be shown that a person has violated it. As with all criminal cases, prosecutors must prove all elements of a case beyond a reasonable doubt. This is sometimes hard to do when the harasser's motives are unclear. Thus in 2006, a Pennsylvania woman had her convictions overturned under a statute that made it unlawful "to intentionally obstruct or interfere with the lawful taking of wildlife."[10] The woman, Janice Haagensen, escaped conviction by demonstrating that her intent was to keep people from trespassing on her lands. She also sought to halt hunters who were violating state game laws by hunting too close to public roads or in oversized groups. This evidence undercut the prosecutor's claim that her primary intent was to obstruct lawful hunting. Ms. Haagensen's defense succeeded even when she was mistaken about land boundaries and harassed hunters on neighboring lands.

CAPTURE ON PRIVATE LAND

Pierson v. Post was an unusual case because the dispute arose on land that was, as the court described it, wild, uninhabited, and unpossessed. We do not know whether the land had a private owner. According to one historical study, the land belonged to the local town. Some local residents but not others had legal rights to use it. In any event, the owner of the wild New York beach made no attempt to assert rights in the captured fox. Land ownership was thus irrelevant.

The court's reference to the land's being "wild" and "unpossessed" rarely draws the attention that it should. What law students are not told, and indeed what the legal community has mostly forgotten, is that American hunters in 1805 largely had free range of the rural countryside to pursue game, regardless of ownership boundaries. Hunting in rural landscapes was one of the liberties of American citizens, one of the key ways in which America differed from Britain. That liberty and the spirit behind it were aptly expressed in a judicial decision handed down in South Carolina in 1818, thirteen years after *Pierson v. Post.*

The South Carolina dispute involved a rural landowner who ordered a deer hunter on horseback to stay off his land. When the hunter continued the hunt, the landowner filed suit, claiming trespass. In the resulting ruling, *M'Conico v. Singleton,* the court roundly rejected the landowner's trespass claim.[11] Indeed, it did so with an air of incredulity. The court expressed surprise that a landowner would even think of asserting a right to keep hunters away, so entrenched was the public liberty of hunting. "Until the bringing of this action [that is, the lawsuit in trespass]," Justice William Johnson explained,

> the right to hunt on unenclosed and uncultivated lands has never been disputed, and it is well known that it has been universally exercised from the first settlement of the country up to the present time; and the time as been, when, in all probability, obedient as our ancestors were to the laws of the country, a civil war would have been the consequence of an attempt, even by the legislature, to enforce a restraint on this privilege. It was the source from whence a great portion of them derived food and raiment, and was, to the devoted huntsman, (disreputable as the life now is,) a source of considerable profit. The forest was regarded as a common, in which they entered at pleasure, and exercised the privilege; and it will not be denied that animals, *feræ naturæ,* are common property, and belong to the first taker. If, therefore, usage can make law, none was ever better established.

Members of the public, in short, had "the right to hunt on unenclosed land." And it was a right not at all dependent on the landowner's permission. Indeed, asserted the court, "it never entered the mind of any man, that a right which the law gives"—that is, the right to hunt on unenclosed land—could "be defeated at the mere will and caprice" of a landowner.

To read this ruling of the South Carolina Supreme Court from 1818 is to be transported to a far different era, to a time when the word "liberty" meant something much different than the landowner's right to keep outsiders away. The ownership of unenclosed land in antebellum America did not give landowners anything like the right to exclude. The public could use such land, and they could do so, in fact, for more than hunting. As the court noted in passing, "the forest being a common," citizens had the right to allow their cattle "to range at large." The grass itself, even on private land, was also a common, free for anyone to take.

This South Carolina opinion gives hints, though, that legal change was already in the offing. Even as the South Carolina court strongly affirmed the practice of keeping rural lands open for hunting, the dispute provided an omen of things to come. At least one rural landowner—the plaintiff in the case—believed that the public's hunting rights should come to an end. We need only to note a parenthetical comment the court made about people who lived off the land, hunting for food—a life perhaps once honorable that, by 1818, had apparently become "disreputable." In frontier days, nearly everyone hunted—all save the maligned gentry, who insisted that others do their work. But that time was largely gone. By 1818, solid Carolina citizens, slaves excluded, tilled their own land and drew sustenance from it in that manner. It was only a matter of time before courts began showing less solicitude for roaming hunters.

The idea that governed early America was that citizens had free use of all unenclosed lands, even when privately owned and without regard for the landowner's wishes. In the case of hunting, colonial charters and state constitutions expressly put this individual right in writing. William Penn's 1683 Frame of Government of Pennsylvania, for instance, protected the liberty of all citizens "to fowl and hunt upon the land they hold, and all other lands therein not enclosed."[12] By "other lands," the constitution clearly meant lands that were owned either by other people or by the colony itself. This language was borrowed nearly verbatim by lawmakers elsewhere. In its constitution, Vermont similarly protected the right of all inhabitants "to hunt and fowl on the lands they hold, and on other lands not enclosed."[13] This right in Vermont also included the right to fish on all "boatable waters," without regard to ownership rights.

Plentiful evidence suggests that the countryside was open in much if not all of the United States as the nineteenth century began. Traveling Englishman John Woods reported on the situation in southeastern Illinois around 1820, highlighting how different it was from the situation back home:

> The time for sporting lasts from the 1st of January to the last day of December, as every person has a right of sporting, on all unenclosed land, for all sorts of wild animals and game, without any license or qualifications as to property. . . . Many of the Americans will hardly credit you,

> if you inform them, there is any country in the world where one order of men are allowed to kill and eat game, to the exclusion of all others. But when you tell then that the occupiers of land are frequently among this number, they lose all patience, and declare, they would not submit to be so imposed on.[14]

Historian John Mack Faragher testifies to the same situation in central Illinois not long thereafter:

> Sugar Creek [a region near present-day Springfield] farmers, like their ancestors and counterparts throughout the nation, utilized important rural productive resources in common with their neighbors. Custom allowed farmers, for example, to hunt game for their own use though they might be in the woodlands owned by someone else. Hogs running wild in the timber and surviving on the mast paid no heed to property lines. And despite an 1831 prohibition against "stealing" timber from unclaimed congress land, settlers acted as if the resource of these acres belonged to the neighborhood in common and helped themselves, "hooking" whatever timber they needed.[15]

Particularly engaging evidence on rural hunting comes from the famous memoir of William Elliott, *Carolina Sports by Land and Water,* first published in 1846.[16] Writing as "Venator" and "Piscator," Elliott regaled readers with his exploits of devil-fishing and wildcat hunting. He could see, though, that wild game was declining along the Atlantic coast and that the era of the open range, there at least, would soon have to end. The main cause of game disappearance, Elliott reported, was loss of necessary wildlife habitat, particularly deforestation and increased cattle grazing in woods. A further cause was the rise of market hunters, who served hotels and the "private tables of luxurious citizens." Elliott confirmed that "the right to hunt wild animals" was "held by the great body of the people, whether landholders or otherwise, as one of their franchises." The practical effect of this right, he explained, was that a man's rural land was "no longer his, (except in a qualified sense) unless he encloses it. In other respects, it is his neighbors' or any bodys.'" The public could graze animals at will and harry an owner's livestock with hunting dogs. So entrenched was the public's right to hunt, Elliott reported, that some people desired "to extend it to enclosed lands, unconditionally,—or, at least, maintain their right to pursue the game thereon, when started without the enclosure." Even when lands were enclosed, owners had trouble halting public users. Proof of trespass was hard to present, juries were "exceedingly benevolent," and the "the penalty insufficient to deter from a repetition of the offence."

As state economies developed and as fewer and fewer people embraced subsistence modes of hunting, pasturing, and gathering, it was perhaps inevitable that the landowning class would challenge this entrenched public liberty. States such as Vermont, where the right to hunt was emblazoned on the constitution, would keep the public right alive well into the twentieth century. In other states, however, judges began slow marches, step-by-step, toward giving landowners the power to keep hunters away, whether or not their lands were enclosed. Legislatures got into the act as well. New statutes gave landowners the power to exclude public hunters simply by posting conspicuous signs. Nearly everywhere, we should note, the presumption for generations remained that unenclosed lands were open to public hunting. But times were changing. If they chose to do so, landowners could take control.

What the law was doing, at varying speeds in the various states, was giving landowners the power to exclude hunters so long as they were willing to take the affirmative steps required to exercise their legal power. Many states made it difficult for landowners to exercise that right. They insisted that landowners post signs in the precise manner prescribed by law. Any false step would allow hunters to come on. Some states also required landowners to file papers with local government offices, announcing their plans. On their part, many judges continued to resist this legal trend. In countless rulings, local judges would decide, for instance, that No Trespassing signs did not mean "no hunting," or that signs were not adequately visible or in the right places. Well into the twentieth century, many judges were hunters themselves, and placed high value on the activity. Still, the trend seemed inexorable, particularly when the industrial farm lobby got involved and strongly sided with the landowning interest.

Today, the law gives to landowners nearly everywhere the right to exclude public hunters, even from unenclosed lands and even, apparently, in states such as Vermont, where the right to hunt unenclosed land is guaranteed in the state constitution. These days, "unenclosed land" is assumed to mean land that has not been posted, although the phrase when originated certainly did not have that meaning. The Vermont provision, however, continues to inspire challenges to limits on hunting within the state. In 2006, the state supreme court pushed aside a claim by a hunting and trapping association that the constitutional rights of its members were violated when a regional park district banned hunting on 2,000 acres of unenclosed parkland.[17]

Some states have now gone to the farthest end of this long road away from free public hunting. They now provide expressly that lands are closed to public hunting unless the hunter obtains the landowner's permission. More frequent among states is the practice of telling landowners that they can exercise their right to exclude only if they take the steps specified by statute to post their lands against hunting. In such states, the presumption remains that unenclosed rural lands are open for hunting unless the landowner demonstrates otherwise. Many states now have narrowly tailored statutes that permit entry by a hunter in limited circumstances, usually to retrieve animals that have been wounded. In addition, some states distinguish among types of lands in their trespass laws, granting greater protection in the case of lands devoted to agriculture. Perhaps the most prohunter states (Vermont among them) are the few that allow landowners to post their lands against trespassers but then withhold from such landowners certain economic benefits that they might otherwise receive. In Vermont, landowners who post give up the right to recover reimbursement for damages to crops, fruit trees, and crop-bearing plants caused by deer and black bears.

Trespass, as the term is used at law, is both a civil wrong and a crime. Civil wrongs—or what the law generally terms "torts"—can be rectified only if the injured party brings a civil action in court to obtain relief. Crimes, in contrast, are prosecuted by government lawyers. Crimes bring criminal penalties imposed by the state, either fines or imprisonment. In civil actions for trespass, two quite different remedies are available: landowners can recover monetary damages for actual injury done to their land, and, in appropriate cases, they may obtain from a court an order (injunction) commanding a specific person to avoid trespassing in the future.

It's useful to consider these remedies and how they are obtained, because they influence the ways the law really works in practice. Local prosecutors exercise their own discretion as to whether to bring a criminal trespass action; no one can tell them what cases to take. Often overburdened, they typically will prosecute only important criminal violations. Trespass on land is unlikely to be deemed important enough to warrant attention absent some overt injury to the landowner or the land. More significantly, many prosecutors are apparently unwilling to bring a criminal trespass action unless the defendant has been personally warned to stay away or has become a specific nuisance to the landowner. A mere one-time unconsented entry is not likely to warrant prosecution. With respect to civil actions, as noted, it is up to the landowner personally to file the suit and thus pay the legal and court fees. Unless the trespasser has caused significant physical damage to the land, the landowner is unlikely to recover more than a very modest amount as damages—perhaps only a dollar to acknowledge the landowner's rights. A landowner can easily end up losing money after paying all costs. The bottom line is that remedies for minor trespasses, not resulting in damage and not involving a specific person who has become a nuisance, are relatively problematic. This practical reality is an important part of the legal arrangement between landowners and hunters.

Trespassing hunters, of course, sometimes kill game while trespassing. This killing adds a new wrinkle to a trespass dispute. What happens to the animal that the trespasser has killed? The long-confused law in England on this issue was not resolved until long after the American colonies broke away. The authoritative ruling, mentioned in chapter 2, was the 1865 House of Lords decision in *Blades v. Higgs.* There the Lords decided that an animal wrongfully killed by a trespasser belonged to the landowner. The same legal rule now applies in the United States, although the decision in *Blades* did initially draw sharp cultural criticism from some American judges. The tendency today is to rationalize this outcome in pragmatic terms rather than technical legal ones. One policy rationale, used by the Louisiana Supreme Court in 1926, rests on the idea that a trespasser is a wrongdoer and should not benefit from his wrongdoing.[18] As the court explained, the trespassing hunter "must account to [the land] owner for all the fruits of his unlawful exercise . . . this being in accord with the moral maxim of the land that 'no man ought to enrich himself at the expense of another.'" A related, perhaps stronger policy rationale is that the rule in *Blades* tends to discourage trespass and thus deters future wrongdoing. A trespasser who cannot keep what he shoots is less likely to trespass, or so the reasoning goes.

Blades involved an animal that was killed while the hunter was trespassing. A much different case is presented when an animal is mortally wounded in a place where the hunter is *not* trespassing, with the animal then escaping to die on privately owned land. Can the hunter enter the private land to retrieve the animal? The answer varies among states. A few states specifically authorize entry under such circumstances. Entry to retrieve the animal would also be permitted on unposted land in states where posting is required. Regardless of the rule on entry, the animal belongs to the hunter in all states because the mortal wound qualifies as capture of the animal. Thus a mortally wounded animal that enters the private land is, at the time of entry, the property of the hunter and remains the hunter's property even after entry. If the hunter cannot immediately enter to retrieve the animal, then the hunter's remedy is to contact the landowner and ask either for express permission to enter or for the return of the animal. This factual pattern might be likened to the errant baseball that flies into a neighbor's yard. The owner of the baseball continues to own it and can ask for its return. Should the neighbor not return the baseball, the ball owner could sue for its monetary value (in a legal action termed "conversion"). In both instances—the baseball and the wounded animal—the landowner can properly demand compensation for any damage done or for any expenses incurred in retrieving the item.

CAPTURE AND THE GAME LAWS

In all states, the rule of capture is significantly limited in its application by fish and game laws. Such laws limit who can capture, what can be captured, and when and where capture can take place. Licenses are typically required, and they are given out using allocation methods that mix first-in-time with other factors, including age, citizenship, and the payment of fees. Game laws typically apply in full, or nearly so, to landowners who want to hunt their own lands. Thus landowners are distinctly limited in their ability to reduce animals to possession whenever they might like, just as in Britain at the time of the Revolution.

Game laws in the United States were often underenforced throughout much of the nineteenth century. Only after the Civil War was pressure brought to bear on states to tighten their laws and begin enforcing them. Much of this early game conservation work was undertaken by wealthy, often socially elite sportsmen in eastern cities concerned about disappearing game. Some of them went beyond political lobbying to spend their own money hiring people to identify game law violators and to present evidence to prosecutors. A practice in New York City, for instance, was for wealthy sportsmen to send hired agents to city restaurants, to see if the restaurants were serving game caught out of season. Many of them were, and with apparently no inhibitions. The better restaurants, at least, were shamed into stopping the practice.

Given this spotty enforcement record, it is not surprising that legal disputes involving game law violations rarely appeared in courts until late in the nineteenth century. When they did begin appearing, one of the first issues to arise was whether a hunter who took game unlawfully could nonetheless keep it—a variant of the question whether a trespassing hunter could keep game. What became the leading American case on the issue arose in Maine in 1889. A landowner-hunter captured a live moose out of season, took it to his land, and secured it. A game warden found out about the moose, entered the land without permission, and seized the moose, all without any particular legal authority to do so. The landowner then sued to regain the moose. The case was a difficult one for the court because the game warden lacked any special authority; he stood in the same shoes, legally, as any other citizen. Nonetheless, the court ruled that the landowner could not recover the moose because the moose's capture had violated game laws. As the court saw matters, a property right could not arise out of unlawful conduct, whether by a trespasser or by a landowner.

The court's ruling in the case, *James v. Wood,* is particularly interesting because the court presented various factual hypotheticals to illustrate the legal rule that it was announcing.[19] A person who captured an animal in violation of game laws, particularly a live animal, acquired no property interest in it. That was the basic rule. It therefore made sense that, if another person released the captured animal, the wrongful hunter could claim no damages. Holding no property interest, the hunter had lost nothing. The court expressed its reasoning rhetorically:

> Suppose lobsters illegally taken are thrown overboard alive. Is he who does it a trespasser? Shall the taker of them have damages for his illegal catch? Or suppose one lands a salmon in violation of law, and a bystander, while it is yet alive, throws it back into the water. Shall the fisherman have the value of the salmon that the law forbids his having at all?

The court's answers to its rhetorical questions were clear: illegal capture did not create any property rights that the law would protect.

The issue in *James v. Wood* resurfaced in a 2000 federal appellate court ruling from Rhode Island, *Bilida v. McCleod,* involving a pet raccoon that a local government seized because the owner lacked a required permit.[20] The alleged owner brought an action against the government, claiming that her property had been taken in violation of her constitutional rights. The court rejected the claim on the ground that the plaintiff had no property rights in the raccoon. The raccoon continued to be owned by Rhode Island in its sovereign capacity. For that reason, the government took nothing when it seized the raccoon (or, more aptly, the government simply took possession of what it still owned).

James v. Wood and *Bilida v. McCleod* dealt with the unusual case of an animal captured alive. The more typical case has involved unlawful killing of wildlife in violation of game laws. In such cases, courts have also said that wrongful hunters and fishers can have their game seized by the state. When strictly applied, this legal rule can take hunters by surprise. In a Montana Supreme Court case from 1968, for instance, elk hunters had their game seized because the guide whom they employed was not properly licensed, even though the hunters were.[21] The court also ruled that the state could confiscate lawfully killed game if game tags are not properly filled out and affixed to the animals. State fish and game codes now routinely include provisions allowing agents to seize animals that are unlawfully taken. Such laws violate no property rights for the reason mentioned: a hunter acquires no property rights when an animal is unlawfully taken and thus loses nothing when the animal is seized.

THE DURATION AND EXTENT OF PRIVATE RIGHTS

Once an animal is lawfully acquired, with no trespass and no violation of game laws involved in the capture, what property rights does the capturer acquire? How do property rights in wildlife differ from rights owners have in other property?

From an early date, Anglo-American law embraced what presumably had long been a customary understanding since time immemorial, that a wild animal that escapes back to the wild becomes available for capture by someone else. Property rights in live animals, that is, are limited or "qualified" in duration; they last only so long as the animal is under control. A common exception to this rule is that an owner retains rights in an escaping animal so long as the owner actively pursues the animal with the intent to retrieve it.

A number of judicial rulings in the nineteenth and early twentieth centuries applied these basic principles to swarms of bees, which necessarily require freedom of movement. The rule that gained support was that an owner of a swarm of bees retains ownership of it, even when it escapes, so long as the owner spots the swarm and actively pursues it. As a New York court in 1836 put it, "If a swarm fly from the hive of another, his qualified property continues in them so long as he can keep them in sight, and possesses the power to pursue them."[22] That ownership continues even if the bees alight on the private land of someone else. One such case, involving a bee owner who was a lawyer, led a judge in the City Court of Yonkers, New York, in 1916 to lace his ruling with a bit of humor:

> It is not the intention of the court to review the history of bees, of their origin in Asia, and of how they have followed man in the development of civilization, and have been tamed and used by him, and have contributed to his comfort and welfare, although it is an interesting study and one to be recommended. We shall pass without comment the fact that the claimant to these bees is a lawyer, and that no lawyer needs bees to assist him in stinging. The thing really turns on the question of identity [of the particular bees], and at the preliminary hearing, when the courtroom was nearly filled with lawyers, I offered to appoint any one of them a referee to make personal examination of these bees for marks of identification, which offer they unanimously rejected, the first time in my experience with the law that a lawyer has refused to accept a reference.[23]

An animal owner, in sum, retains rights in an escaping animal so long as the owner actively pursues it and does not give up the search. Another wrinkle in this law has to do with animals that by their nature wander off and then return, individual bees among them. Here, too, the rule has been that an owner retains rights in animals when and so long as they exhibit a reliable habit of returning.

A more difficult question has had to do with the requirement that an escaping animal, to regain liberty, must return to something like its natural habitat. A circus elephant that escapes on a city street does not thereby become unowned; it belongs to the circus, whether or not it is actively pursued. This means both that the circus can retrieve it and that the circus is liable as owner for any damage the elephant does. A New York case from 1898, *Mullett v. Bradley,* highlights the problem that can arise in cases less clear than the wandering circus elephant.[24] In *Mullett,* the plaintiff captured a sea lion near San Francisco and shipped it to New York. There it escaped into the Atlantic Ocean and was subsequently recaptured. The original owner sought to retrieve it, arguing that sea lions were native only to the Pacific Ocean and did not naturally live anywhere near New York City. The court rejected the argument because the animal in fact was living and flourishing in the wild. The question, the court said, was whether the animal had regained its "natural liberty." That happened when, "by its own volition, it has escaped from all artificial restraint, and is free to follow the bent of its natural inclinations."

If we can rely upon the few common law rulings that have been handed down over the years, it appears that an animal need not return to its native habitat in order to regain its liberty. But if it need not, then how do we distinguish the sea lion case from that of the escaping zoo elephant? And what rule applies to escaping parakeets and monkeys in Florida, parrots in San Francisco, and the many pet snakes and reptiles that have escaped or been released?

As for animals deliberately released, the owner has abandoned them. Abandonment based on an owner's intent to give up rights automatically ends the owner's private rights. There is no need to decide whether wildlife law on its own would have terminated the property rights. As for an escaped animal that the owner wants to recover, perhaps the key policy question is whether someone spotting the animal would realize that it is privately owned. That issue, in turn, seems to have two parts to it: is the animal one that an owner might have abandoned, and could the animal simply be a wild creature living in the area, never before owned? When a person who spots the roaming animal exerts considerable energy capturing it, it seems unfair to take the animal away, claiming that it still belongs to the prior owner, unless the subsequent capturer had fair notice that a prior owner existed.

These various policy considerations help clarify the law. An elephant is obviously not the kind of animal that would be abandoned; it is simply too valuable. It is also not native to any part of America. No sensible person would think otherwise. Clearly, anyone capturing an elephant would know that it is owned by someone somewhere. As for the sea lion, a good many people would not realize that it naturally lives nowhere near New York City. Much the same can be said for the various parakeets, snakes, ferrets, and other animals that people hold as pets. They, too, might appear native to many parts of the country. In any event, people often release unwanted pets into the wild. Because many pets can thrive in the wild, even escaping pets would likely be deemed unowned unless well marked. Whether a subsequent person can capture such animals, however, raises other questions. Escaped pets that regain their status as animals living in the wild become subject to state wildlife laws. Parakeets, for instance, might well qualify as songbirds and thus be protected under state, if not federal, law.

An interesting quirk of the law is the old rule that "base" animals could not be the subject of the crime of larceny. A thief could not be prosecuted for stealing them. The original rationale for the rule apparently was that such animals lacked moral value, but it was sustained into modern times by the more plausible rationale that such animals simply had no monetary value. The rule reflected the sensible idea that a person shouldn't go to jail for the theft of a thing of no value. Perhaps the legal rule also had a humanitarian rationale: a poor person who stole and killed a base animal for lack of other food should be pitied, not punished.

For generations, this odd legal rule has drawn regular ridicule from courts. The North Carolina Supreme Court in 1871, for instance, was asked to apply the rule by a defendant charged with theft of an otter from a trap.[25] The otter would have been considered a base animal, but the court thought its economic value to trappers was more significant than its moral status:

> All of the distinctions as to animals *feræ naturæ* as to their generous or base natures, which we find in the English books, will not hold good in this country. . . . We take the true criterion to be, the *value* of the animals, whether for the food of man, for its fur, or otherwise.

Similarly, the Supreme Court of Arkansas in 1883 considered it silly for a thief to claim that, while he might be prosecuted for the theft of a birdcage, he could not be prosecuted for the theft of the "base" mockingbird in it, which its sentimental owner deemed "priceless."[26]

A further limit on the property rights a person holds in wildlife, perhaps the most significant of them, is the continuing power of government to redefine property rights in ways that disadvantage the owner. Governments have inherent powers over property generally. This state power is often termed a "regulatory power"; at other times it's described as the power to redefine property rights. Governments are apparently more inclined to exercise this power in the case of wildlife than with other types of property. When they do exercise it, the result is more likely to curtail private rights seriously.

An illustration of this possibility, prominent because the case rose to the United States Supreme Court, had to do with a federal ban on selling eagle feathers.[27] Eagles were endangered largely due to the commercial trade in feathers and other eagle parts and, later, due to pesticide use practices. Lawmakers responded by enacting the Bald Eagle Protection Act of 1940,[28] which included a sweeping regulatory ban on the sale of all feathers, without regard for when and how the feathers were obtained. Plaintiffs in the case held stocks of lawfully obtained feathers, intending to sell them. When they could no longer do so, the feathers lost their market value. The Supreme Court in 1979 nonetheless upheld the validity of the regulations against a legal claim that they effectively took private property unconstitutionally without the payment of compensation. According to the Supreme Court in *Andrus v. Allard,* a prohibition on the sale of lawfully acquired property is a legitimate lawmaking activity when the prohibition is reasonably grounded in policy.

Many cities and counties ban residents from possessing various animals, either because the animals are thought dangerous or because of disease risks. Such laws have routinely withstood challenges, even when applied to animals that people already possess. States have also succeeded in defending new laws that ban rural landowners from possessing various categories of exotic animals, including animals awaiting sale to zoos or game ranches. Such laws usually take effect at some point in the future, giving owners a chance to transport the animals for sale out of state. But surrounding states may have similar laws of their own, including bans on importing particular species. As a result, animal owners can find themselves unable to sell their animals and barred even from continuing to possess them.

A 1915 ruling by the Washington Supreme Court, *Graves v Dunlap,* gives a good look both at the legal risks animal owners face and at the solicitude that courts occasionally show to protect owners from a new law's harshness.[29] The case involved a dairy farmer who had taken in two deer and raised them as pets. The deer grew to a herd of twenty animals. The owner had also captured pheasants and raised them. While the dairy farmer held the animals, new game laws took effect, banning the possession, killing, and sale of game out of season. Both the deer and the pheasants qualified as game. The owner challenged the new statutes as applied to his unusual circumstances, especially the ban on possession out of season. The Washington court upheld the validity of the entire law. At the same time, it crafted an exception for the dairy farmer. The farmer could keep the animals out of season, the court ruled. He could also kill any diseased or injured animal as a mercy killing. Otherwise the owner had to abide in full by the laws prohibiting killing and sale out of season.

Graves v. Dunlap achieved a fair result that did not undermine law enforcement efforts by the state. The dairy owner, however, got an unusually fortunate result; many courts would not have been so solicitous. No doubt the ruling was influenced by the fact that the owner treated the animals as pets, not as inventory or breeding stock. Bans on possessing animals have often gone into effect and been applied to animals already possessed. Owners have no recourse but to comply, either by killing the animals or selling them quickly.

THE SPECIAL PROBLEMS OF GAME RANCHES

These basic common law rules of animal ownership have been stretched and challenged in recent decades due to the many game ranches and commercial operations that raise wild animals in fenced areas. What is the status of such animals, particularly when they roam freely over large spaces and live as if fully wild? Only a few judicial decisions have dealt directly with game operations. Still, the annals of the law provide useful guidance, both on the rules of law that are likely to govern and on the legal issues that legislatures could usefully clarify.

In thinking about game ranches, we can return to the initial issue of animal capture. What happens when a landowner constructs a high, sturdy fence over dozens, hundreds, or even thousands of acres of land? Has the landowner thereby captured the animals that are fenced in? Obviously they haven't been taken into physical possession, animal by animal. Indeed, the landowner might not know they exist. Should that make a difference legally?

Our immediate assumption could be that landowners prefer to hold property rights in the animals. After all, why not own something? But this prediction might be wrong. Many of the animals that a landowner "captures" could be protected by game laws, with criminal penalties imposed for their unlawful capture. Other animals that are fenced in might be on the game list, but with the landowner guilty of taking too many or taking them out of season. Then there are the various liability rules that may be triggered if the enclosed animals cause harm to visitors. As we'll see in chapter 7, landowners are not liable for harm caused by wild animals, only by animals that are owned. In short, the landowner might have many reasons why she *doesn't* want to claim that fenced-in animals become her private property.

What, then, is the legal status of these animals?

A rare case in which this issue arose was *State v. Brogan,* a Montana ruling from 1993.[30] A landowner was convicted criminally for entrapping wild elk when he opened the gates on his farm pastures and baited the pastures with hay. In the course of upholding the criminal conviction, the court commented that the landowner also did *not* gain ownership rights in the many deer that had wandered into the fenced pastures and that were inadvertently trapped when the fences were closed.

Aside from *Brogan,* judicial rulings involving commercial operations have mostly had to do either with captive-bred animals or with animals that have been captured and then intentionally released into enclosed areas. The rule of capture is not an issue in such instances because the animals have already been captured. What, then, about the legal status of animals once they are released onto private land? Are they still privately owned if they can roam freely over large areas? How large must an area be to qualify as natural habitat so that a released animal is deemed to have escaped to the wild? And what about the presence or absence of fences?

Before answering these questions, it's useful to consider what is at stake legally, aside from rules governing liability for wildlife-caused harm. One body of law implicated is state criminal law. Criminal law protects people from having their property stolen or intentionally destroyed. If animals in fenced areas are viewed as private property, then a trespassing poacher could be arrested for stealing or destroying private property, not just taking game out of season. One of the dreaded evils of old England was the tendency to lock up poachers for theft, or even, on occasion, put them to death. That odious practice was condemned in early America as the height of aristocratic arrogance. Are we on the verge of embracing the practice ourselves?

This issue of fenced animals, and whether they have or have not regained their natural liberties, becomes more confusing when we consider animals that are not mobile. If a fence is adequate to keep roaming animals in private ownership, then is a fence even needed for animals that cannot leave an area or for animals that are unlikely, by temperament and behavior, to do so? Must a game rancher, that is, construct a fence for such animals if the fence is not really needed to keep them in place? Medieval English wildlife law dealt separately with animals that lacked the physical ability to escape, often very young animals. They were said to be adequately under private control simply because they could not physically escape. That legal category has largely been lost to legal memory. Yet it lurks in legal history, perhaps to rise again.

For help in probing these game ranch questions we can turn to old judicial rulings involving fish and shellfish. One useful opinion was handed down in 1893 by the Maryland Supreme Court, in the case of *Sollers v. Sollers.*[31] There a fisherman fenced in a natural cove in the Patuxent River. After catching fish, he placed them alive in the fenced-in area of the cove. When an interloper took the fish, the fisherman sued in trespass. The court ruled against the fisherman. As the court saw things, the fisherman lost his property rights in the fish when they were put in the cove. They had sufficiently regained their liberty to become unowned.

Also of interest are judicial rulings dealing with commercial oyster operations. Early lawsuits involved oyster owners who placed their oysters in public waters to propagate with the intent of harvesting the offspring. The oysters did not move, of course, and the young oysters mostly remained nearby. The fishers were confident that few of their animals would get away. Still, the oysters were in their natural habitat, and it was difficult for other fishers to recognize that the oysters were privately claimed. Disputes of this type arose frequently and with conflicting legal outcomes. Some courts emphasized that the oysters were sedentary and had no realistic chance to escape; they therefore remained privately owned. Other courts stressed the public's rights in the waters and the difficulties outsiders had of knowing when mollusks were privately claimed. These courts tended to say that the oysters had escaped and become unowned, thus available for capture by someone else. The law remains fragmented.

Early mollusk cases were especially complex for two other legal reasons. One was that courts well into the nineteenth century tended to allow private fishers to stake off public waters. By using them exclusively for a number of years, they could claim exclusive fishing rights in them. That practice came to an end later in the century: courts decided that navigable waters should remain open to public fishing, free of any claims of exclusive private rights. While it lasted, though, this legal rule allowed users of oyster beds to assert special legal rights to keep using them. The second complicating factor was that oysters and other mollusks attached themselves to the submerged land, and the land itself was often privately owned. Courts then had to decide whether mollusks attached to land were part of the land itself, like a tree, thus giving the landowner the exclusive right to capture them, or whether they were more like swimming fish, available for public fishers to catch.

The obvious solution in these mollusk disputes—the one that several states finally embraced—was to set up state commissions to regulate oyster beds and lease rights to particular fishers. The common law, standing alone, just wasn't adequate to produce a sensible regulatory regime for the beds. The same lesson might well apply to today's game ranches. Legislatures need to step in to make rules governing the ranches, paying particular attention to the property rights, if any, that game ranchers have in their free-ranging animals and to the many liability issues that no doubt will come up.

One final case worth mentioning, related to the problems of game ranches, is the celebrated 1927 decision of the Colorado Supreme Court in *E. A. Stephens and Co. v. Albers*, involving a silver fox named McKenzie Duncan.[32] Silver foxes were highly valued in the 1920s and bred for their fur. Born and raised in captivity, McKenzie Duncan was identified by an ear tattoo. Duncan escaped from his cage and soon fell victim to the shotgun of a rancher, who discovered him prowling near his chicken coop. Knowing nothing about the fox industry, the rancher sold the pelt to a trapper. The trapper, in turn, sold it to a fur company agent, who apparently realized he was buying the pelt of a fox raised in captivity. From the pellet holes in the pelt the buyer also knew the animal had not been killed by the original owner, who would not have mutilated a valuable pelt.

The Colorado court's ruling is instructive because the court ruminated on the lack of useful law to resolve such disputes. The common law rule about escaping wild animals didn't seem to fit when animals were part of a ranching operation. While the dispute about Duncan was pending, the Colorado legislature took up the policy issue, under lobbying pressure from the state's fur-producing industry. The legislature enacted an extremely impractical law stating that captive-raised fur-bearing animals and their offspring forever (!) would be considered domesticated animals. Under this statute, owners of domesticated foxes retained their rights in the foxes long after escape, generation upon generation. As the court noted, this new statute made no sense. What happened when escaped animals mated with wild ones? How were people to distinguish domesticated animals from wild ones when they roamed at large? And what about trappers in the wild who unknowingly trapped domesticated animals? Would they be liable for unknowingly capturing a privately owned fox?

The Colorado court ultimately resolved the dispute over McKenzie Duncan based on the knowledge of the fur company agent who bought the pelt. The pelt had pellet holes in it, and the buyer knew it was captive bred. The buyer had adequate notice that it was privately claimed and therefore had to return the pelt. One wonders, though, what the court would have done if the less-knowledgeable rancher had still possessed the pelt, the one who shot the fox because it was preying upon his chickens? And what would have happened if the original owner of the animal had sued for its value as a brood stock—as a living animal—rather than just as a pelt? Surely the rancher would not be liable in money damages for protecting his chickens? It also doesn't make sense to hold the pelt buyer legally responsible for the death of the fox, since it was dead when he bought it.

The full legal status of game ranches cannot be pieced together until we deal with issues in chapters 4 and 7 addressing wildlife on private land and the rules of liability that go along with owning wild animals. But already we can see problems dealing with the rule of capture, the rule of escape, and the general issues about the kinds of property rights that a person has in wild animals. A thoughtful judicial commentary on this suite of issues was offered by Judge Tom Rickhoff of the Texas Court of Appeals in a concurring opinion in a 1994 case, *State v. Bartee.*[33] In that case, the court held that Texas could not prosecute hunters for theft when they killed deer out of season. Texas, to be sure, had sovereign ownership of the animals. But this sovereignty was inadequate to sustain a conviction for stealing state property. According to the court, deer could be stolen as a matter of criminal law only when held in captivity as private property.

In his separate opinion filed in the case, Judge Rickhoff considered the plight of deer held on private ranches or otherwise lawfully confined. Could they be objects of theft when confined to an area hundreds of acres in size, he asked rhetorically? The state plainly held the power to regulate or prohibit the fencing of large areas. But once a state allowed fences, Rickhoff claimed, "much of the responsibility for preserving and increasing our wildlife [will have] fallen to the owners of these fenced preserves." Having made this observation, Rickhoff added an intriguing comment: "With that responsibility should also come the rights, duties and liabilities of ownership."

As readers of Rickhoff's opinion, we are left to wonder: What might those "rights, duties and liabilities of ownership" entail? Would a landowner be liable if escaping animals cause harm to people or property? What if the animal harbors and spreads disease to domesticated livestock? Also, could a landowner allow people to enter the land and "harvest" the privately owned animals—just like domestic livestock—without regard for hunting laws and license requirements? If so, what would this mean for game officials, who already have trouble enough enforcing state fish and game laws?

Some states have begun to address these issues in their laws and regulations. But their full resolution, chiefly by legislative and regulatory action, has not yet come. No doubt more years of experience with game ranches and some legislative trial and error will help inform future lawmaking.

CHAPTER

4

Wildlife on Private Land

With 60 percent of the United States in private hands, a great deal of the nation's wildlife lives on privately owned land. A major challenge of wildlife law has been to establish the legal relationship between the private owner of land and publicly owned wildlife. We have already seen some of the principles of wildlife law that apply to private land. Other pieces, though, are required to complete the legal picture. What rights do landowners themselves have in such wildlife? What legal protections do landowners enjoy when engaged in wildlife-related activities? What can they do when wildlife causes harm? And, finally, what legal issues arise when landowners allow outsiders to hunt on their lands?

As noted in chapter 2, landowners have considerable power to exclude public hunters. They also have the right to claim game that is taken by trespassers, at least if the state does not seize it because of game law violations. On its side, the state wields vast power to prohibit hunting on private land. That power includes at least some ability to keep landowners from degrading wildlife habitat, though the extent of that state power has not yet been tested.

Before we turn to the issues of this chapter, it is helpful to consider a few fundamentals about private property rights generally. Private rights in land, like all property rights, derive from laws enacted by government. In the case of land and wildlife, state law largely controls, with federal and local law playing complementary roles. Although private property is often viewed as an individual right, something akin to free speech or freedom of religion, it is in fact a quite different entitlement. An obvious difference is that citizens have free speech rights simply by being alive. In the case of property, in contrast, citizens have no rights unless they buy property or otherwise acquire it. An equally significant difference is that private property is not a right that arises out of the Constitution, as free speech does. The Constitution only safeguards property rights that have been authorized by other bodies of law, chiefly state law.

What all this means is that a property owner enjoys protection only for property rights that are recognized by statutes or by the common law. As the law changes over time, so do the property rights based upon them. The government, then, has broad powers to redefine and regulate private rights in terms of what an owner can and cannot do. In fact, it makes good sense for property to work this way as a matter of political legitimacy. Private property gives to an owner the power to call upon the police to arrest trespassers, thus depriving citizens of their liberties to roam. This arrangement—this legal curtailment of the liberties of nonowners—is morally legitimate only if it promotes the good of society generally. It is not enough, to justify this use of state power to coerce nonowners, to show that property benefits its owner. It must be shown that the arrangement benefits society generally, including the nonowners. Should it appear that property law gives an owner power to cause harm in some way, then the time may have come for the law to change—as it has many times over the generations.

It is useful to keep these basic points in mind when thinking about wildlife. They particularly help in appreciating the considerable tension that surrounds wildlife on private land. On the one hand, landowners often want power to do whatever they like on their lands, no matter how harmful it is to wildlife. If land includes valuable habitat, owners may want the power to destroy that habitat, to make way for farming, housing, or commerce. When habitat is destroyed, of course, wildlife can suffer greatly, the wildlife that is owned by the people collectively. As co-owners, the people might have quite a different desire—for example, to have at least some of the habitat protected. And thus the conflict. We might phrase this conflict in several ways. Should the desires of the landowner consistently take precedence, as they typically have, or should the law provide some protection for wildlife in recognition of the public's property rights in them? Similarly, when have a landowner's activities shifted across the longstanding line, from socially beneficially to socially harmful, triggering a need to impose limits on what landowners can do?

Such philosophic questions may seem distant from the day-to-day problems of wildlife management. But they underlie the complex of legal issues that arise when publicly owned animals live on privately owned land. As we look into the future, no element of wildlife law is likely to experience greater stress. Particularly in the case of rare species, conservation needs to take place on private land. And it needs to take place in a way that is fair to everyone while respecting private property as a vital public institution. Fairness needs to include fairness to taxpayers, who, it would seem, shouldn't have to pay people to halt land uses that undercut the common good.

A basic element of landownership for centuries has been the principle "do no harm." Landowners have never had the legal right to undertake activities that cause harm, either to neighbors or the surrounding community. A key challenge for lawmakers in years to come will be to tailor that basic idea to the plight of wildlife on private land. How much freedom should landowners have to alter their lands in ways that harm wildlife? How intensively should they be able to use their lands before violating the do-no-harm rule? On the other hand, when has society imposed too heavy a conservation burden on particular landowners so that society itself, rather than the landowners, should help shoulder the cost?

THE ALLEGED RIGHT TO HUNT

The power of states to protect wildlife includes the power to ban hunting and other wildlife takings. Bans can apply statewide or operate more selectively. This regulatory power, of course, can annoy landowners. It can leave them unable to hunt on their own lands. In the case of people who bought land chiefly for hunting, or who paid for the right to hunt someone else's land, such laws can be particularly unsettling. But the simple fact is that the vast majority of landowners in the United States cannot hunt on their land since hunting is generally prohibited in residential areas.

Probably most bothersome are regulations that allow some landowners to hunt but bar others from doing so. Laws of that type appear to treat landowners unequally. They seem to run afoul of one of our nation's most cherished rights, the right to equal treatment.

The typical view of courts on the matter of hunting bans is illustrated by a ruling from the Supreme Court of Washington in 1914. In the case *Cawsey v. Brickey,* hunters formed a gun club and leased land on which to hunt.[1] They apparently constructed a club building and otherwise spent "considerable sums" on their hunting activity. In 1913, Washington enacted a statute (long since repealed) that authorized county game commissions to designate particular lands as game preserves, where no hunting would take place. The sole limit was that commissioners could designate as preserves no more than three townships within each county. The lands leased by the gun club were located within a designated game preserve. Because its hunting activities were frustrated, the gun club challenged the ban.

According to the Washington court, the state plainly had the power to prohibit hunting on private land. Game on private land was owned by the state, and landowners had no power to hunt or capture it except as authorized by law. The landowner had only a qualified right or, as the court put it, "more correctly speaking," merely "a privilege," to keep trespassing hunters off his property. The landowner had no rights in the game itself. Indeed, in relation to the state, the landowner had no enforceable property rights whatsoever.

What made the case complex, as the court sized it up, was the fact that some private lands were covered by the hunting ban while others were not. Did the commissioners' action, therefore, "bear unequally on persons similarly situated so as to be obnoxious to the constitutional inhibition against class legislation?" Did it violate the fundamental idea of equal treatment? In the court's view, it did not. The law, the court asserted, treated all people alike *as people:* no person could hunt in a no-hunting zone, and any licensed hunter could hunt in places open to hunting. Thus all people were subject to the same rules. What the law did instead was to distinguish among parcels of *land.* This was entirely proper, the court said, because the lands themselves differed considerably in terms of habitat value. Lands that included valuable habitat were not the same as lands that lacked good habitat. There was no reason why the law couldn't consider these differences:

> The owner of land which from its location and character is peculiarly suited for a game preserve is not situated similarly to other landowners with reference to the subject-matter and purpose of a law creating a preserve. The subject-matter and purpose is protection and preservation of

> game. It is so declared in the title of the act. One whose land is thus peculiarly suited to meet those purposes obviously occupies a different relation to the purpose of the law from that occupied by one whose land is not so suited.

In its ruling, the Washington court distinguished this gun club dispute from the facts of a Florida ruling handed down four years earlier, in 1910. In the Florida case, residents of one county were given an exemption from a statewide law under which anyone wanting to hunt in a given area had "to give three days' notice to the game warden and pay a special license tax."[2] That law distinguished among people based on residence, not among land parcels, and the residence of hunters had no bearing on any conservation aim. The law thus treated people unequally, violating the state's constitution.

Cawsey v. Brickey represents the near-universal view among states. Landowners have no right to hunt that is legally enforceable against the state. They therefore cannot complain simply because the state bans or restricts hunting. As the Wisconsin Supreme Court put it in 1962, "It is . . . well settled that hunting is a privilege as against the state (commonly called hunting rights in reference to land), which the state can grant, deny, or regulate."[3] In *Collopy v. Wildlife Commission,* the Colorado Supreme Court in 1981 considered a landowner's claim that a ban on hunting within a four-square-mile area was unconstitutional.[4] According to the disgruntled landowner, the ban allegedly amounted to a "taking" without compensation of the landowner's "independent common law property right to hunt such game upon his own land." The court roundly rejected the argument. The flaw in the landowner's argument, the court ruled, was that the landowner had no such property right. "We here decide that the right to hunt wild game upon one's own land is not a property right enforceable against the state."

A more narrowly tailored argument was put before a federal appellate court in 1995 by landowners in *Clayjon Production Corp. v. Petera.*[5] While the case was pending, it drew considerable attention within the "property rights" movement, then enjoying public visibility. At trial, the landowners argued that they had "the exclusive right to hunt the wild animals that [were] present on their property." Because of that ownership, they contended, the state's entire wildlife regulatory system was invalid as applied to private land. On appeal, after their argument was rejected at trial, the landowners presented a new, more modest argument. What landowners possessed, they claimed, was a property right in the "harvestable surplus" of wildlife on their lands. That is, they allegedly owned the animals produced on their lands that were in excess of those needed to keep breeding populations in good shape. The court also rejected this more refined argument, although it based its ruling on constitutional grounds without passing judgment on the landowners' novel property rights claim.

The idea that a landowner has a "right to hunt" enforceable against the state has been recognized in only one appellate ruling, the 1963 opinion of the Florida Supreme Court in *Alford v. Finch.*[6] Although the ruling is not a good indication of wildlife law today, it is nonetheless useful to consider. *Alford* involved a suspicious conservation plan implemented by the state's Game and Freshwater Fish Commission. Under the plan, lands of certain landowners were opened to hunting while all surrounding lands were closed. The plan was worked out by the state commission in private negotiations with the landowners whose lands would be open to hunting. Landowners whose lands were closed were not consulted. Understandably, the Florida Supreme Court reacted negatively to the closed-door arrangement. While agreeing that the state owned all wildlife in its sovereign capacity, the court decided that the state had unlawfully deprived landowners of their rights "to pursue game" upon their lands.

Alford is as an aberration among judicial rulings. Other courts have expressly discredited it. In *Alford* the Florida court showed confusion about the common law of wildlife. It also erred in the precedents that it cited for its ruling. (It cited cases recognizing the powers of landowners to transfer hunting easements in their lands—an entirely different legal issue, taken up later in this chapter.) Still, *Alford* is usefully remembered because it shows how courts can strive to protect landowners when they sense that state agencies are unfairly discriminating among them. Courts may be particularly prone to do so when a conservation plan contains a distinct hint of cronyism. The fact that the arrangement in *Alford* was formulated behind closed doors, through negotiations only with the landowners receiving favorable treatment, no doubt added to the court's suspicion.

Overall, then, it is clear that states have the power to draw lines on the map, allowing hunting in some places and not others. They merely must have a reasonable basis for doing so. This legal conclusion assumes, however, that the line drawing is done by a legal entity with power to do it. In a 1965 ruling, *Allen v. McClellan,* the New Mexico Supreme Court struck down a state agency's attempt to designate particular lands as a game preserve. The reason: the agency's charter from the state did not give it the power to designate preserves.[7]

DAMAGE CAUSED BY WILDLIFE

Wildlife on private land can sometimes cause damage, particularly to growing crops and livestock. Wildlife damage typically shows up in the courtroom in two distinct ways. First, a landowner may sue the state for compensation for the harm that has occurred. The argument typically used in such cases is that the state should be liable for the harm because the animals causing it were publicly owned. Second, a landowner may claim that he has the right to defend his property by killing wildlife in violation of game laws. This second claim is normally raised as a defense when the landowner is prosecuted for a game law violation.

The law in this area is rather easy to summarize. Landowners are able to get compensation from the state for wildlife damage only when the state expressly authorizes it. The state's ownership of wildlife is insufficient to make the state liable for harm without its consent. This rule of law was applied in *Barrett v. State,* the 1917 New York ruling about beaver damage considered in chapter 2. It has consistently governed ever since. For instance, a federal appellate court in a 1950 ruling, *Sickman v. United States,* decided that the federal government was not liable for damage done by geese protected under the Migratory Bird Treaty Act of 1918.[8] In a 1984 ruling, the Alaska appellate court decided that a landowner had no claim when a protected bear entered his land and destroyed a moose carcass.[9] A California appellate court in 1994 ruled that the state was not liable for the destruction of fences and loss of forage caused by reintroduced elk.[10]

These various property damage cases need to be distinguished from cases in which wild animals cause *bodily* injury, a legal issue considered in chapter 7. As we shall see there, government can be liable for personal injury caused by wildlife when the injury is closely tied to the negligence of government agents.

Many states today voluntarily offer compensation for particular types of wildlife damage, so long as landowners properly document the harm and act within time periods specified. Statutes usually compensate for damage to agricultural crops and to livestock attacked by protected predators.

Compensation statutes often draw fine distinctions in terms of who can collect and when, with courts following the laws carefully. Thus in a 1993 ruling, the Wyoming Supreme Court denied recovery for livestock damage caused by bison.[11] The statute authorized compensation only for damage done "by big or trophy game animals or game birds." Bison were not a game animal and thus were not covered by the statute.

Landowners seeking to defend their property against wildlife have had somewhat more success in arguing that they can do so, despite game laws. Here too, however, courts typically side with the state. Landowners usually succeed only when they've taken all reasonable steps to protect their property against the wildlife and when they have pursued all avenues for getting state help. An illustrative case is *Cross v. State,* a 1962 Wyoming decision.[12] The landowner in that dispute was charged with killing moose out of season and without a license. The defendant and his family owned seventy-two hundred acres of land. As wildlife populations rose over several decades, they suffered regular damage totaling several thousand dollars per year, chiefly the loss of forage, damage to fencing, and disruptions of livestock operations. For years, the defendant chased the wild elk and moose off his land, only to have them return. He also

> hired and paid for riders to get the wild animals driven away; he has expended considerable sums of money hiring airplanes to "spook" the moose and elk away from his premises; he has been forced into long and drawn out litigation with the Game and Fish Department over the years in an effort to induce the department to enforce sufficient control to protect the residents in this areas; this defendant and various of his neighbors have been forced to support and maintain this ever-increasing herd of wild game species and have been helpless to prevent the wild game from virtually "taking over" the ranching operations by belligerently driving cattle from feed grounds, by chasing horses, by chasing hired help and their families indoors, by usurping and defending their usurpation of sheds, barns, hay corrals, etc.

The defendant was charged with violating game laws when he killed two moose. He had been unable to drive the two moose away by "spooking" them, and the moose had, in his presence, become entangled in a wire fence, ripping out "considerable good fence."

On the facts of the case, the court ruled that the landowner could assert a defense-of-property claim to his criminal prosecution. The court summarized this defense and explained when it could be used:

> Before a defendant can resort to force in protecting his property from wild animals protected by law he should use every remedy available to him before killing such animals. He should use only such force as may be reasonably necessary and suitable to protect his property and must use only such force and means as a reasonably prudent man would use under the circumstances.

In a few other reported decisions, landowners have also succeeded in avoiding conviction by raising a defense-of-property claim. The common practice, as in *Cross,* has been to require a defendant to show diligence in using all other methods to protect the property, including working with state officials. In a 2008 ruling, the Washington Supreme Court went a bit further in recognizing a landowner's right to defend property, in this instance defending orchards from predation by elk.[13] The court held that the owner could take action so long as it was reasonably necessary, with necessity judged in part by the availability of other steps the landowner might take—requesting action by the state, obtaining advance permission to kill the animals, and seeking compensation after the damage has occurred. Occasionally, a court has assisted defendants by deciding that a particular game law does not apply to animals causing severe damage. In *Cook v. State,* from 1937, the Washington Supreme Court decided that the state statute governing the trapping of fur-bearing animals did not apply to animals that were destroying a private dam.[14]

Despite these successes, the general practice today is to deny landowners any right to harm marauding wildlife unless state law expressly allows it. Thus the New Mexico Supreme Court in 1999 refused to allow a criminal defendant to raise this defense when he killed deer that were destroying crops and pasture.[15] The defense was unavailable even though the deer problem had gone on for twenty years, the deer herd sometimes numbered one hundred animals, and the landowner had worked for years with the state game and fish department to alleviate the problem. In a 2001 ruling, the Idaho appellate court similarly rejected the defense when put forth by a landowner who shot a deer that was eating garden plants.[16] In a 1988 ruling, a federal court of appeals decided that the owner of grazing rights on federal land had no right to kill a grizzly bear that repeatedly raided his flock of sheep.[17]

Over the past two decades, many states have enacted statutes authorizing landowners to kill marauding animals under prescribed circumstances. These statutes vary in terms of which animals can be taken and based on what types of damage. Some states require landowners to notify state officials in advance and condition the killing of an animal on the exhaustion of nonlethal control efforts. Some states require landowners to obtain licenses; others specify that animals can be killed only if "found in the act" of doing damage. Many states require that landowners report a killing to the state, sometimes within short time periods. Connecticut prescribes that licensed animal control businesses may kill animals only using methods that "conform to the recommendations of the 1993 report of the Veterinary Medical Association panel on euthanasia."[18]

These statutes need to be read along with more general wildlife laws that specify which animals are protected. Many annoying animals are, in fact, unprotected at any time. The typical state arrangement is that wildlife falls into three basic categories: game animals, which can be killed only in accordance with game laws; protected animals (usually the vast majority of all species), which cannot be killed at any time; and unprotected animals, which include the common law category of noxious or destructive species known as "vermin." The practice in many states is to identify the specific types of animals that enjoy no legal protection. These animals might be listed directly, or they may be designated indirectly by being left off lists of protected species. Thus common species of mice and rats, for instance, typically enjoy no protection. Among birds, starlings and house sparrows can often be freely killed. Landowners, however, need to be careful before assuming that a species enjoys no protection. Common pigeons, for instance, are often covered by protective laws. Mammals such as skunks and opossums may or may not be.

A final issue worth raising, in terms of the landowner's power to protect against wildlife, has to do with constructing fences to exclude wildlife. This typically unobjectionable practice becomes troubling when fences interfere with critical patterns of animal migration or movement. When fences do cause harm, their legitimacy quickly comes into question. If a landowner cannot kill an animal directly, even when it is causing harm, should the landowner be able to do so indirectly by excluding the animal in a way that causes death?

Several fencing disputes have yielded interesting judicial rulings in recent years. All have upheld the power of government to prohibit fences. A 1995 Florida ruling upheld the validity of a law prohibiting fences that interfered with an endangered species of deer.[19] Similarly, a New York ruling from 2000 upheld a state order directing a landowner to remove a thirty-five-hundred-foot-long "snake-proof" fence that blocked migration of a threatened snake species.[20] Perhaps most interesting has been the federal appellate court ruling in *United States ex rel. Bergen v. Lawrence* from 1988.[21] There a cattle rancher constructed a twenty-eight-mile fence, enclosing his private ranch along with nearly ten thousand acres of public grazing lands to protect it against migrating pronghorn. In the harsh winter of 1983, the fence kept pronghorn from reaching their winter feeding grounds. Many of them starved. According to the court, the fence violated the federal Unlawful Enclosures Act, which prohibits fences that enclose federally owned land. An alternative argument was that the landowner's fence unlawfully obstructed an implied easement that allowed pronghorn to migrate across the private land. The implied easement argument, although novel, was viewed by the court as potentially meritorious. In the end, the court resolved the case without deciding whether such an easement existed.

HABITAT MODIFICATION

Running through all these decisions—whether dealing with game preserves, limits on harming wildlife and recovering damages, or limits on the right to exclude—is a fundamental question of social policy. What rights should landowners have to use their lands in ways that disrupt wildlife populations? What limits might society reasonable impose on them? Because private property is an evolving institution with legal rules that gradually shift over time, yesterday's answers may not be the same as tomorrow's. Weighty considerations lie on both sides of the habitat issue.

We can use as illustration of this issue the case of the sheep rancher who is troubled by predation. Sheep can be raised in most parts of the country with no serious predator problems or with problems easily solved with guard animals. With wool supplies so abundant and mutton bringing low prices, what need does the nation have to promote sheep ranching in areas where sheep and predators come into conflict? The nation could easily meet its needs for mutton and wool without raising sheep in places where wolves and bears pose problems. That being so, what social values are promoted by favoring sheep when they come into conflict with rare predators? Landowners, of course, want flexibility to use their lands as they like. This flexibility is a form of individual freedom that America rightly takes seriously. But in the case of sheep in bear or wolf country, how important is this abstract liberty interest? Could we not say to sheep ranchers what we already tell a wide range of other economic producers: find a place for your activity that is well suited for it. Industries can hardly set up wherever they choose; they must find appropriate places. Land uses in urban areas are regularly zoned so they occur in places best suited for them. Should the same rule apply to rural land uses such as sheep ranching?

Such questions are not easy to answer, and, again, tomorrow's answers may differ from today's. One important trend now taking place in land use law generally can shed light on this dispute. In many specific ways, the law is slowly redefining landowner rights so that the rights again reflect the conclusion that each parcel of land is unique. In part, this uniqueness reflects the land's natural features. That is, what a landowner can do on the land is increasingly depending on the land's natural features and on the ecological effects that a given land use will have. This was the idea sketched in *Cawsey v. Brickey*, the ruling by the Washington Supreme Court that rested on the idea that different laws could apply to different types of natural lands. The idea now applies widely. For instance, landowners have only limited rights to dredge, fill in, or drain various types of wetlands. More strict are laws prohibiting construction in floodplains or on fragile barrier islands. Farmers in many states are limited in their ability to remove vegetation along stream banks or to plow sloping fields in ways that erode soil. These and many similar laws all deal with narrow conservation issues. But, when pieced together, they give evidence of a more general trend toward tailoring landowner rights to the land's natural characteristics. The basic idea is that landowners ought to avoid land uses that entail altering the land in ecologically harmful ways. Such reasoning might seem novel in rural areas, but it is very much in line with the thinking that has governed urban land planning for generations.

In the case of wildlife law, the critical question has to do with wildlife habitat and the ability of landowners to alter it as a prelude to initiating some other land use. Few courts or lawmakers have addressed the issue directly. When we start searching, however, we find that there already is a wide range of laws that restrict the ability of landowners to alter habitat. Laws protecting wetlands and floodplains offer examples. Laws controlling polluted water runoff can also force landowners indirectly to protect native vegetation, including native wildlife habitat. Forestry practices statutes in various states limit the ability of large forest owners to clear-cut timber; they, too, can protect wildlife habitat. Other legal examples could also be cited.

One decision worth particular mention is the Washington appellate court's 1980 ruling in *State Department of Fisheries v. Gillette.*[22] The dispute involved a farmer whose land bordered a salmon spawning stream. Flooding along the stream resulted in severe stream bank erosion, so much so that a utility pole was left dangling unsupported along the stream edge. To deal with the problem, the landowner brought in earthmoving equipment to rebuild the stream bank and alter the stream. The landowner, though, failed to comply with the state law requiring the written approval of both the director of fisheries and the director of game before interfering with any streambed. The landowner was penalized for failing to do so. Of particular interest was the fact that the state department of fisheries sought, as an additional remedy, an award of money damages compensating the state for the loss of the salmon fishery. The award was based on the state's sovereign ownership of the fish. The court upheld the damage award, noting that the state had both the power to act and "the fiduciary obligation of any trustee to seek damages for injury to the object of its trust."

Gillette is an unusual decision in the sense that there are few similar decisions. Few other fish kill cases have been based on the physical modification of wildlife habitat, as opposed to chemical contamination. Moreover, the landowner's activity in the case directly violated a state statute. Without the statute, the landowner might well have the power to do what he did. Still, the decision deserves study because it could prove an early sign of legal change. The landowner had no intention of harming the salmon fishery, but his actions nonetheless did so. In the case of salmon fisheries, the public continues to spend large sums of money modifying dams and otherwise attempting to promote healthy salmon populations. A chief reason why salmon populations remain low is because of habitat modification of the type involved in *Gillette.* If salmon streams need better care, how should it come about? Should taxpayers be expected to pay to halt degrading activities, or is it reasonable for lawmakers to demand that landowners take better care of the streams?

One setting in which habitat modification has been restricted legally is the case of habitat critical to the survival of rare species. At the federal level (as we shall see in later chapters), the Endangered Species Act of 1973 bars landowners from altering designated critical habitat when it actually kills or injures threatened or endangered animals. Landowners can go ahead with their projects only if they get federal permits to do so. Permits are available only if landowners prepare habitat conservation plans. This law has been politically contentious, drawing sharp criticism from libertarian and prodeveloper interests. But it has withstood challenge and seems likely to stay in place. Many state laws have similar provisions protecting the habitats of state-listed species.

One way to think about such laws and where they might be heading is to compare them with the longstanding rule of land ownership, noted above, that owners must avoid acting in ways that cause harm. Property law has a long history, with many twists and turns, in the ways it specifies how landowners can use their lands. One lesson from this history is that each generation has seen fit to define land use "harm" as it wants. Harm, that is, is a flexible notion. It depends for its precise meaning upon widely prevailing ideas about what land uses are proper. Early in the nineteenth century, many landowners were prohibited from constructing water-powered mills because of the fish they killed. As the century went on, the law tilted to favor industrial development. Many landowners got away with mining activities and other land uses that resulted in massive fish kills. And they did so because popular sentiment favored industrial development.

As we look forward, it is hard to predict in what direction property law will head. It is difficult to know what types of land use activities will be deemed harmful. Will the law shift back to something like its form two hundred years ago and offer greater protection for wildlife and for subsistence wildlife-harvesting activities? Or will the law continue to embrace ideas crafted during the proindustry decades of the nineteenth century, the ideas of property as an abstract, uniform concept that favor developers and industrial land users?

By way of concluding this quick look ahead, we might take a backward glance to England in the reign of William and Mary. In 1692, Parliament enacted a statute to protect wildlife habitat from activities that harmed its habitat value:

> Be it enacted, That for the better preserving the red and black game of grouse . . . no person whatsoever, on any mountains, hills, heaths, moors, forests, chases, or other wastes, shall presume to burn, between

> the second day of February and twenty-fourth of June, any grig, ling, heath, furze, goss, or fern, upon pain that the offender or offenders shall be committed to the house of correction, for any time not exceeding one month and not less than ten days, there to be whipt, and kept to hard labour.[23]

As far back as the reign of Richard II in 1393, statutes similarly required builders of dams to include weirs "of reasonable wideness" to permit spawning runs to pass through their dams.[24] The lesson here might be summarized as follows: when lawmakers have viewed particular wildlife species as valuable, they have taken steps to protect their habitats.

PROTECTIONS FOR WILDLIFE-RELATED ACTIVITIES

Many animals can be sources of value for landowners, both monetarily and otherwise. Predictably, one of wildlife law's functions has been to give legal protection to landowners who engage in wildlife-related activities, particularly activities that generate economic returns.

A classic English wildlife law case from 1707, *Keeble v. Hickeringill*, illustrates how this sometimes happens. A landowner erected on his pond various "decoy ducks, nets, machines, and other engines for decoying and taking wildfowl."[25] A neighboring landowner, desirous of disrupting the duck-harvesting effort, repeatedly frightened the ducks away by discharging "six guns laden with gunpowder." The court ruled that the pond owner had been harmed, and awarded him compensation for his lost income. Had the neighbor simply sought to attract the ducks himself—that is, had he merely engaged in fair competition in the race to capture the ducks—his activities would have been acceptable. But the neighbor's illegitimate aim was instead to disrupt the capture. He thus became liable for the lost revenue to the duck harvester. The duck harvester, it is important to note, had not yet captured the ducks and thus did not own them. The legal injury in the case, therefore, was not the theft or destruction of privately owned property. Instead, it was the unlawful disturbance of a legitimate business enterprise.

Landowners in America have often recovered monetary damages for harm to their wildlife activities. One form of relief is for trespass onto land that results in killing wildlife or otherwise disturbing a wildlife activity. Thus in a Louisiana appellate ruling from 1955, *Harrison v. Petroleum Surveys,*

Inc., a landowner who raised muskrats on his marshlands sued a petroleum surveying company for physically damaging his operation.[26] The survey company, confused about property boundaries, entered the plaintiff's land in marsh buggies—"huge, heavy vehicles which churn through the marsh . . . leaving tracks from 18 inches to four feet deep, depending on the softness of the marsh, and approximately twelve feet in width." The marsh buggies "completely crushed" two acres of the plaintiff's land and the marsh grass growing on it. The two-acre tract was prime habitat for raising muskrats. It was home to an estimated six hundred to a thousand muskrats, including young, and produced approximately one hundred muskrats per year. Several hundred of the muskrats were directly killed by the trespassing buggies. The marsh was expected to take some eight to ten years to recover from the damage.

The defendant's argument in *Harrison* was that the landowner could not recover for the muskrats because they were animals living in the wild, not in confinement. The landowner therefore did not own them. The court agreed: the landowner did not own the muskrats and thus could not recover monetarily for the animals directly killed. The landowner, however, could recover for the lost income from the muskrat operation, which was a legitimate use of the land. In calculating this lost income, the court decided that the plaintiff would lose one hundred muskrats per year for an eight-year period, with the profits operation totaling fifty-seven cents per muskrat. Had the trespass been "intentional or willful or wanton," the landowner might have recovered punitive damages as well.

Harrison involved a straightforward case of trespass that did physical damage to the land. To prove a trespass, a landowner must show a direct, physical invasion of her land that was undertaken without consent. Many times the harm to the wildlife operation is less direct, as in the English case of *Keeble v. Hickeringill.* Indirect harms to wildlife are chiefly covered by the common law of nuisance. Nuisance law allows a landowner to recover for harm caused to her use and enjoyment of land or for interferences with such uses. To recover, the landowner must show essentially two things: that the interference produced "substantial" harm, and that the imposition of the harm under the circumstances was unreasonable. Nuisance is a broad, catchall category that covers harms not involving a direct, physical invasion of the land (which is covered in trespass). Nuisance is harder for a landowner to prove because these two elements are often difficult to satisfy. A trespass can be shown even though the harm is minimal. In nuisance, the plaintiff must show that the economic or physical harm was substantial. And in trespass, any unconsented physical invasion will suffice; in the case of nuisance the interference must in some sense be unreasonable. *Keeble v. Hickeringill* would today be termed a nuisance case.

An instructive wildlife damage case was handed down in 1949 by the California Court of Appeals in *Lenk v. Spezia.*[27] The case was brought by a beekeeper who lost fourteen tons of honey due to the death of his bees. The bees were killed when they flew onto nearby farm fields that had been sprayed with toxic insecticides to protect farm crops. Because there was no evidence that the insecticides were sprayed onto the plaintiff's land, the case did not involve trespass. Instead, the beekeeper brought the case as a negligence action, on the ground that the farmer failed to use due care in his operations.

On the facts, the insecticide sprayer regularly notified the beekeeper in advance of the spraying and asked him to move the hives or to keep the bees temporarily confined. The beekeeper refused to do so, even though fully aware of the risks to the bees. Had the farmer been negligent in his spraying, the court agreed, he would be liable for damage caused. Liability would occur also if the pesticides spread onto the plaintiff's land. Here, however, the sprayer used caution. Also important, the court said, was that the bees were in effect trespassing on the farmer's property. Because the animals were trespassing, the farmer owed a much lesser level of care to guard against harming them. On the facts, the plaintiff beekeeper failed to provide enough evidence that the harm was caused by negligence.

Landowners, then, have a range of common law tort actions that they can bring to protect their wildlife-related activities, including trespass, nuisance, and negligence. As these cases illustrate, though, landowners cannot recover directly for the loss of animals they do not own. Their recovery is based instead on the loss of income from their activities or upon other harm they have suffered.

Looking ahead, one wonders what courts will say when they confront wildlife activity cases in which landowners are not interested in killing the animals (as in *Keeble* and *Harrison*) or putting them to work (as in *Lenk*), but instead are merely interested in watching them. The traditional description of the interest protected by a nuisance action is the land occupier's "use and enjoyment" of the land. One of the highest-valued uses of wildlife today is the enjoyment people get from watching and studying them. Wildlife can be of obvious economic value to owners of resorts or other vacation properties who use the wildlife to draw visitors. But even ordinary landowners draw enjoyment from wildlife, as attested by the millions of birdwatchers and backyard bird feeders. Then there are the landowners who devote their land chiefly to wildlife habitat as an intentional conservation activity. These land uses, too, should enjoy protection.

The problem courts will face is not in deciding whether these land uses deserve protection; they plainly do. It will be in deciding how the basic rules of trespass, nuisance, and negligence apply to them. How would damages be calculated? And when might a landowner obtain an injunction to halt activities that are seriously disturbing the wildlife? The questions are not easy ones, and courts will likely have difficulty with them. It is typically said that the law does not protect unusually sensitive land use activities. But this generalization, like the do-no-harm generalization, does not really tell us much until we decide what land uses are deemed sensitive and which are not. In an age when industrialization ruled supreme, many land uses were deemed too sensitive to warrant protection. But what might the law say tomorrow, if society decides that it wants to live more in harmony with nature and to protect quiet, settled modes of living?

HUNTING EASEMENTS

A final topic involves the transaction commonly known as a hunting easement—that is, the transaction in which a landowner grants formal permission to another person to come onto his land to hunt. With the decline of free public hunting as landowners prohibit access to their lands, arrangements of this type have become increasingly common.

Hunting easements and related hunting arrangements are not particularly complex in legal terms. They raise the same legal questions as do all types of easements and land use licenses. What differs are the peculiar facts involved and the need for easements to cover the full range of specific problems that can arise between hunters and landowners. Too many easement agreements are brief documents, covering only a few of the pertinent issues. It is only later, when problems crop up, that the parties wish they had spent more time hammering out details and drafting an agreement that specified the respective rights clearly.

Permission to enter land to hunt can be granted in many ways and can be subject to almost an infinite variety of terms, from the casual oral permission to enter for a day to a full-blown conveyance of perpetual, exclusive rights to engage in all lawful hunting. The practical issues that arise when this is done—the issues that a well-written hunting agreement would cover—are rather easy to list: How long does the right last? Who can enter the land, and when and where? Are there limits to the numbers of hunters who can come on, and how long can they stay? What animals can be taken, and how? Can the holder of the right sell it or otherwise transfer it to someone else? Can the right be divided, in the sense that the holder can sell the same right to multiple people? Is any rental or other fee involved? Can the landowner unilaterally cancel the right or change its fundamental terms in any way? When can the hunter complain about land use practices by the owner that diminish hunting opportunities? And what happens if game laws suddenly ban hunting on the land?

Sometimes hunting arrangements envision that the holder of the hunting right can construct a cabin, deer stand or other structure on the land. Obviously, the parties' understanding on this issue, whatever it is, should be set forth clearly. Also, what happens when the hunting agreement ends? Can the holder of it remove any structure that's been built? Is he instead entitled to compensation if the structure adds value to the property and is left behind? Again, the more detail in the agreement, the better.

For the most part, the parties to a hunting agreement are free to write it as they see fit. That is, they can resolve the many questions in whatever way they like, and courts will enforce the terms of their agreement. The law's main role is to provide back-up terms to deal with gaps in contracts—that is, any issues and possible problems that the parties failed to include in their agreement. Needless to say, it is useful for landowners and hunters to know these backup rules. If the parties are content with the terms set by the law, then they need not deal with that issue in their agreement. If they don't like the law's backup rules, they need to take steps to replace them.

Without question, the most important issue in hunting agreements is whether an agreement qualifies as an "easement" or whether it is merely what the law terms a "license." The two categories differ fundamentally at law. It is important that parties decide which arrangement they want to create. An *easement* is an interest in land that is perpetual in duration unless otherwise agreed. An easement can be freely bought and sold. Once granted, the landowner has no power to cancel it or to alter its terms unless the power is expressly reserved. The easement is binding, not just on the landowner who granted it, but on all subsequent owners of the land (at least if the easement is recorded or subsequent owners have notice of it). Unless the easement states otherwise, a landowner cannot even cancel the easement based on the easement holder's violations of its terms.

In sharp contrast to an easement is a *license,* which is not viewed as a property interest and which is much less durable. A license can be either a contract right, if obtained under a binding contract, or merely casual permission to do something. A license generally cannot be transferred unless the parties decide otherwise; in fact, an attempt to transfer a license may bring it to an end. A license is also automatically terminated upon the landowner's death or upon sale of the land by the owner. At any time, even if the license is issued pursuant to a binding contract, the landowner can cancel it. If cancellation of the license breaches a contract, the landowner may have to pay damages, but the cancellation remains valid.

Many times parties who draft agreements will use one of these labels, either easement or license, without knowing what they mean. Courts are well aware of this confusion and often ignore the language that parties use. Courts routinely decide for themselves, from the substance of the agreement reached, whether it is properly termed an easement or a license. The more informal, personal, and transient an agreement seems to be, the more likely it is that a court will view it as a license.

Occasionally parties drafting hunting agreements will identify them as "leases." That term is rarely appropriate as a legal matter, since a lease by its nature gives the recipient an exclusive right to use an identified space. Courts routinely ignore the "lease" language in agreements as they go about deciding whether a dispute involves an easement or a license. Lease language is most likely to make sense when it involves a right to construct a building that will be used exclusively by the hunter-lessee.

Whether the parties intend an easement or instead a license, the scope of the hunter's activities ought to be specified with as much detail as possible to avoid surprises. Issues that particularly need addressing include the numbers and perhaps the identities of the people who can come on the property, where on the property they can go, and what liability they might have for any damage done. The use of dogs and horses, if permitted, should of course be covered, as should, today, any questions about advanced hunting equipment and technology. Many landowners prefer that hunters avoid using rifles, particularly long-range ones; issues such as this should be addressed directly.

Aside from these issues, which many people cover in their agreements, a number of basic property-related issues should be given thought. A peculiar feature of easements is that they often "attach to" or are "appurtenant to" a tract of land in the sense that the holder of the easement owns it in her capacity as owner of the land. Thus an easement to cross someone's land to get access to an otherwise land-locked parcel would almost certainly be deemed appurtenant to the land-locked parcel. When the parcel is sold, the easement goes along with it. In the case of hunting easements, the normal presumption is that they do not work this way. An easement to hunt is likely to be deemed the individual property of the person who acquired it. That person is free to hold it or to sell it as an independent item of property. But the situation could be different, particularly when a landowner grants a hunting easement to a neighboring landowner. Is the neighbor acquiring the easement in his capacity as landowner, so that the easement becomes appurtenant to his land? If he is, then a sale of the neighbor's land will sell the hunting easement along with it. The parties to a hunting easement can arrange the easement however they see fit, making it either appurtenant or (as the law terms it) "in gross." As on other issues, though, it is wise for contracting parties to be clear on this point. If they aren't, a court may have to make up its own mind in the course of expensive litigation.

One of the most difficult issues for courts, when they are forced to interpret easement agreements that are incomplete, is to decide whether the easement holder has the right to transfer it to someone else. Generally, easements are fully transferable, which means that a landowner after granting an easement typically has no ability to control the person who ends up using it. But the parties, if they like, can make an agreement nontransferable, or they can give the landowner the power to veto any transfer. Courts often will study an agreement to decide whether the parties implicitly intended to restrain transfers. Particularly when an easement is transferred for little or no consideration, and when the parties are friends or relatives, courts often decide that an easement is personal to the holder. The issue is easily dealt with in an easement agreement, yet it is often overlooked.

Related to the issue of transfer are two other issues: can the holder of the easement divide it, in the sense of granting separate hunting rights to different people, and is the easement meant to be exclusive, so that the landowner can no longer grant hunting rights to other people (and may not even have the right to hunt himself)? Once again, easement law provides backup answers. An easement that is *in gross* (held by one or more hunters personally) is normally presumed to be *nonexclusive*. The landowner can thus sell similar easements to other people, as many as the owner likes. The easement holder, in contrast, can sell the easement but cannot divide it. (This makes sense, because the easement holder would otherwise be competing with the landowner in the sale of hunting rights.) When an easement is *exclusive*, on the other hand, the easement holder rather than the landowner is presumed to have the right to divide the easement, granting hunting rights to multiple people.

This rule on transfer may sound clear enough, but, like many legal rules, it has its ambiguities. What happens when the holder of a nonexclusive easement transfers it, not to some other individual hunter, but to a single hunting group, to be used by the many members of the group? Is that an unlawful attempt to divide the easement, or is it a legitimate exercise of the right to transfer? No easy answer exists, which is to say that those who draft easements ought to address the issue themselves.

As for the issue of how long a hunting right lasts and whether it can be terminated, there exists, as noted, a major difference between easements and licenses. Easements are perpetual unless the parties have decided otherwise. In some states, statutes terminate easements automatically after a number of years (for example, forty years) unless the easement holder rerecords the easement in the county recorder's office, thereby indicating an intent to keep it alive. An easement can also be lost by abandonment—that is, by the holder's relinquishment of it. Technically, abandonment requires an intent to abandon. The mere nonuse of an easement will not lead to abandonment, no matter how long it goes on. Nonuse, however, does provide evidence of an intent to abandon. Abandonment cases are among the most contentious of all easement cases.

A final issue worth considering has to do with the ability of the landowner to change the uses of the land covered by a hunting easement in ways that render it useless for hunting. Is this permitted, and does the easement holder deserve compensation? The law on the subject is not clear. Two rulings over the past few decades highlight the legal uncertainty and thus the wisdom of covering the issue in a written agreement. In one case, an owner of forty-two acres of land wanted to construct a twenty-six-unit condominium project. The land was covered by a hunting and fishing easement granted to a shooting club nearly a century earlier. In a 1994 ruling resolving the dispute, the Wisconsin Supreme Court held that the landowner could not construct the condominiums without violating the property rights of the easement holder.[28] In contrast, the Iowa Supreme Court reached a contrary conclusion in a 1979 ruling.[29] The landowner in that dispute removed the timber on his land, destroyed a pond used for fishing and turtling, and planted the tract in corn. The court upheld the action, stating that a landowner can make free use of the land "absent an express covenant to the contrary or a malicious bad faith destruction of the clearly designated object (i.e., game or its habitat) of the right."

CHAPTER

5

Inland Fisheries

In the history of wildlife law, a sizable majority of all the early disputes dealt with fisheries rather than with game on dry (or "fast") land. Fisheries were economically valuable, and people went to court to fight over them. Fishing cases were also common because fish move around, making it hard to allocate property rights in them or even to mark off exclusive fishing territories. Then there were the many problems that arose when people blocked migrating fish with waterway obstructions, often mill dams to generate power, and when people dumped pollution into rivers. Legal disputes were especially common in states along the Atlantic coast, where fishing and industry were both important.

Rarely did these many lawsuits deal directly with legal rights to the fish themselves. Under the rule of capture, fish are unowned until someone catches them. Instead, the cases dealt with control over the fisheries, which mostly depended upon ownership of the land beneath the waterways. Ownership of submerged land, in turn, had a lot to do (as we'll see) with whether or not a waterway was legally navigable. The public could fish in navigable waterways, while fishing rights in nonnavigable waters belonged exclusively to the owners of the adjacent land. Fishing litigation, accordingly, very often turned on the legal definition of navigability and whether or not particular waterways were navigable under the legal definition. Ironically, almost no cases relating to the navigability of a waterway arose out of a dispute about public rights to use the waterway for travel and transportation—for navigation as we would understand the term. Navigation cases were about fishing.

With all this litigation about fishing rights, it is surprising that the pertinent law on the subject is so unclear today. We would expect that time and litigation would lead to legal clarity. But it hasn't. In many jurisdictions, it remains difficult to figure out which waterbodies are open to public fishing and which are not. Indeed, it is difficult to find a legal issue that is more tangled and uncertain.

This chapter addresses issues dealing with inland fisheries. The principal question has to do with the scope of public rights to use them. Included in that broad topic are questions about, not just *which* waterways are open, but *what rights* the public has in waterbodies that are legally open. Do public rights, for instance, include the right to set traps? to shoot waterfowl? to enter fast land to get around waterway obstructions? A particularly intriguing question has to do with public access to the surface of *nonnavigable* lakes for fishing and boating. Also covered in this chapter are the abilities of landowners to complain when their rights to fish, which are included in their bundle of rights as riparian landowners, are disrupted by people who have polluted a waterway or diminished its water flow. Finally, we'll consider whether and when fishers can complain about interferences with their activities, even though they have no landed property rights and are simply fishing in public waters.

Receiving only slight attention here is an issue that is gaining in importance nationwide: how rights to fish fit together legally with private rights to divert water and apply that water to a beneficial use. Can holders of water rights divert water from rivers when the effect is to harm valuable fisheries that someone else owns? We'll glance at this big issue but will leave it to water law treatises to address fully.

In the intriguing law of fisheries, more so than in any other aspect of wildlife law, it is worthwhile to begin with a solid grounding in the English law upon which American law is based. In the law of fishing, history continues to loom large. Rights to fish in England were complicated by many factors: by claims of royal power, by ancient customary practices, and by disputes about the navigability of waterways. It took centuries for England's highest courts to provide clear answers to even the most basic questions of legal right. One critical issue—what the term "navigability" meant—was not fully resolved until the nineteenth century, after centuries of uncertainty. Interestingly, American judges in the early decades after independence had only spotty access to English law books, some of which were written in the anachronistic language of Law French and largely unintelligible. Without knowing it, early American lawyers were confused about the state of English law. They didn't realize that English law cases embraced two quite different legal definitions of navigability. American judges seized upon one of the definitions, not realizing that there was another.

As it turned out, American judges picked the winner, the definition that English courts ultimately settled upon: that navigability was determined by the ebb and flow of the tides. Only *tidal waters* were navigable. But having picked this definition of navigability, American courts beginning early in the nineteenth century decided to deviate from it. They did so by shifting to what had been the minority strand of thought in England: that waterways that could actually be used for navigation were legally navigable. Navigability, in other words, was a factual question. As American law traveled along this confused and confusing course, it became easy to forget that the foundational issue in this whole area of law was about the rights of the public to fish in waterways, not transportation. Meanwhile, the definition of navigability became important in the United States for legal reasons quite unrelated to fishing and travel: to the allocation of property rights (including the drawing of state boundaries), and to the allocation of decision-making power between the federal and state governments (for instance, the scope of interstate commerce power and admiralty jurisdiction). As these various legal issues were rising up, the United States Supreme Court entered the fray, causing further turmoil. It announced, to the surprise of many observers, that lands beneath navigable waterways were impressed with a special "public trust." This trust obligated states to look after the lands to make sure the public's interests in them were well protected. The Supreme Court's language in time would spawn interest in the idea that a widely applicable public trust doctrine might be used to revitalize natural resources law generally, in ways that better protect ecological health and the public interest.

THE ENGLISH BACKGROUND

Before unraveling this confusion and setting forth as best we can where American law stands today, it will help to summarize our legal inheritance from England. There are seven key elements in the evolution of English common law of navigability and fishing rights, as described below.

Fishing and Land Ownership

The right to fish in English waterways depended upon the ownership of the land beneath the water. Fishing was a legal right attached to land ownership, just like landowners had exclusive rights to cut the trees on their lands.

Landownership and Navigability

The ownership of submerged land was determined in the first instance by the navigability of the waterway. When a waterway was legally navigable, then the underlying land was owned by the king or queen. In contrast, the land underlying nonnavigable waterways was owned by the owner of the adjacent fast lands. If the river was the dividing line between two private owners, each owned to the thread of the river.

Declining Royal Power to Create Exclusive Fisheries

For centuries, the king exercised the power to grant exclusive private rights in waterways to royal favorites, without regard for the navigability of a waterway. These grants were quite similar to the royal grants of exclusive hunting rights on fast land, and at times were just as publicly contentious. Over the centuries, the king's power to make such grants largely faded away as members of Parliament and aristocrats resisted. By the time of American independence, the king's power to make exclusive grants was apparently gone, though royalist writers continued to speak of it and old grants remained in effect.

Royal Ownership Is Sovereign, Not Proprietary

As we noted in chapter 2, the courts decided that the king owned wildlife as a *sovereign* rather than a *proprietary* right. The courts reached the same conclusion on the king's ownership of land under navigable waterways. This meant that the king was supposed to manage fisheries in navigable waters for public use, with the public given free access. An exception was provided for the few species of "royal" fish (whales, sturgeon, dolphins), which were largely viewed as the king's personal property even when found in public waters. The English public made extensive use of its rights to fish in navigable waterways, and fiercely defended them.

Navigability, Defined

This left, then, the looming central issue: which waterways were navigable? For centuries, legal writers debated the issue. Uncertainty lingered because few legal disputes led to authoritative judicial opinions, and these opinions, as noted, provided conflicting interpretations. The conflict was possible because England and, later, Great Britain, did not have a unitary court system with a single Supreme Court as in the United States and in each of our fifty states. Within their respective legal jurisdictions, several courts were supreme. Only the House of Lords sitting as a judicial body could resolve an issue authoritatively. Many British writers and a few courts concluded that navigability was an issue of *fact.* Thus any waterway that was actually used for navigation was navigable, and the public had rights to it. That legal interpretation, however, did not ultimately prevail. Instead, the ascendant interpretation was that navigability was determined by the ebb and flow of the tides. Only *tidal waters* were navigable. Nontidal waters were not, no matter how broad and well used.

With inland travel and transportation so important to England's economy, we might wonder how England's lawmakers could have reached this seemingly odd result. The reason, it turns out, is rather simple. The debates over navigability were limited largely to fishing issues. Most inland waterways that could be used for travel and transport were open to the public for such uses. Indeed, chapter 33 of the Magna Carta of 1215 dealt with that issue, keeping all waterways in England open to travel, just as they were under continental law. Even writers who thought navigability should be limited to tidal waters readily agreed that inland waters used by ordinary people were open to them as of right. Sir Matthew Hale, in his prominent treatise *De Jure Maris* referred to them as "common highways."[1] Others termed them "public highways." Ironically, then, disputes over navigability at law had little or nothing to do with navigation in the sense of travel and transport.

Landownership in Nonnavigable Waters

Because the king's power over navigable waterways ultimately was deemed a sovereign power rather than a proprietary one, and because fishing rights depended upon the ownership of land underneath the waterway, the law of fishing folded back to affect the interpretation of land titles. When a person gained ownership of land adjacent to a *nonnavigable* waterway, his land title went to the center or thread of the waterway. In the case of *navigable* waters, in contrast, grants of land were viewed as conveying only land to the water's edge. Thus land beneath navigable waters remained in the king's hands, subject to sovereign powers of governance and the public's right to fish in them.

Decline of Exclusive Rights in Navigable Waters

As the public's fishing rights in navigable waters became entrenched at law and even more in the popular mind, they called into question the legal legitimacy of prior grants of exclusive fishing rights in such waters. Were these earlier grants of fishing rights still valid if the king really didn't have the legal authority to grant them because he held control of navigable waters in his sovereign, governmental capacity rather than in a proprietary capacity as an individual landowner? Also questionable were longstanding claims made by private citizens that they owned exclusive rights to fish in navigable waterways based on customary understandings where they lived, or based on their long-established exclusive use of a fishery. These adverse-possession-type claims were all based on the presumption that, sometime in the misty past, some ruler of England had granted a royal patent to the fishery to the present owner's ancestors. But the fiction of a long-lost patent to the fishery no longer made sense when the king's power to grant such patents was denied. The pressure built, then, to recover for public use all fisheries in navigable waterways. If that could be done—if all royal patents to fisheries could be extinguished—then the once-complicated law of fisheries would become quite simple: the public could fish in all waterways subject to the ebb and flow of the tides; all other fisheries were exclusively controlled by the private owners of the underlying land, even when the waterway was a public highway for travel and transport.

ENGLISH LAW IN AMERICA

These various principles of English law all crossed the Atlantic, albeit in garbled form. As modified a bit by New World circumstances, they formed the American law of fishing.

American courts in the decades after independence believed that navigability under English law was based on the reach of the tides. Few lawyers were sufficiently versed in English law to realize that an alternative line of legal thought existed in England, linking navigability to the physical suitability of a waterway for public use. With this tidal influence rule to guide them, American courts began to mimic English law as they understood it. English *game* law was highly offensive to many revolutionaries and widely rejected. English *fishing* law, though, was much less so. Perhaps one reason why English fishing law was more acceptable in America was because private fishing rights based on adjacent land ownership were not enforced with much vigor. Landowners who possessed exclusive rights apparently were far from scrupulous in enforcing their rights. In practice, the public could fish at will. Just as the new nation's unenclosed rural lands were open to hunting, grazing, and foraging, so, too, many inland waters were open de facto for public fishing, regardless of legal rights.

American courts, then, instinctively adopted the English rule that fishing rights were based on land ownership. Land ownership, in turn, was based on tidal influences, just as in England. And land beneath tidal waters was held by the government as sovereign, subject to the same public use rights that governed in England. Land beneath nontidal waters, on the other hand, was held by the owners of private adjacent lands, who thus controlled fishing in the waters.

Finally, English law was also followed with respect to the public's right to travel on waterways. Inland waterways were freely open to public use, without regard for legal navigability. Indeed, early courts made it clear that travel on waterways had little or nothing to do with navigability. As a Connecticut court explained in 1818, not just tidal waters, but "all rivers above the flow of the tide, in reference to the use of them, are public, and of consequence, are subservient to the public accommodation. Hence, the fisheries, ferries, bridges, and the internal navigation, are subject to the regulation of government."[2]

Many early judicial rulings in America dealt with the interpretation of colonial land grants, under which the king or a colonial proprietor (such as William Penn) granted private property rights to fish or collect mollusks in tidal waters to particular people. Such grants fomented many legal disputes—continuing, even, to today. From independence, American courts looked upon these royal grants with suspicion, given their taint of royal absolutism. As a New Jersey court reminded readers in an 1812 ruling, an "exclusive right of fishing in the arms of the sea and great rivers, was thought by our English ancestors, a restraint on their natural rights, as odious and more injurious than game laws."[3] Other early disputes that reached the courts had to do with whether private fishers could claim exclusive rights in waterways based on their long exclusive use of them (an adverse-possession-type claim). Several courts at first said, yes, individual fishers could gain exclusive control over parts of navigable waters. But as the nineteenth century went on, courts turned against this idea. By late in that century it was clearly established at law that private parties could *not* gain exclusive rights in public fisheries merely by long and exclusive use.

Hardly had the nineteenth century begun when American courts began to rethink this whole arrangement, beginning with the definition of navigability. The English definition seemed to make little sense given the factual conditions of navigability in America and the vast differences between America and Britain. In Britain, most major rivers flowed directly into the ocean and were subject to tidal influences many miles inland, quite different from the situation in America with its broad inland rivers. One of the first courts to reject the English definition was an 1810 Pennsylvania tribunal, which deemed it rather silly that an important river such as the Susquehanna, "which is a mile wide, . . . and which is navigable and is actually navigated by large boats" would be deemed legally nonnavigable.[4] How could the law be so foolish? Pennsylvania and other courts began what soon became a major trend in American law. They redefined navigability so that it was based upon a river's navigability *in fact*. In practice, as we might guess, that test proved hard to apply, given the huge variation of physical conditions. The physical navigability of a waterway depended on a great many factors, including seasonal water flows, obstructions, and the type of vessel used. Nonetheless, the new legal definition gained influence, with many states adopting it.

In the eastern half of the country, this new test for navigability typically did not change ownership of the beds of rivers that were navigable in fact but not subject to tides. Accordingly, an owner still held title to the center of any nontidal waterway adjacent to his land, even though the waterway was now considered navigable under the new navigability-in-fact test. Surprisingly, this legal shift was apparently made without much thought for its effects on fishing rights. Some states merely assumed that the public now had fishing rights in inland waterways, given the public's common law rights to fish in all navigable waters. Other states, though, pointed to the fact that fishing rights had long been tied to landownership. If the new definition of navigability did not alter property rights in riverbeds (as it didn't), then it logically shouldn't affect fishing rights.

One wonders, reading these early cases, why courts decided to redefine navigability if they did not want to alter public fishing rights. The public, as noted, already had rights to travel on inland rivers not subject to the tides; they were common highways. Something, clearly, was going on, having to do with confusion in the legal mind and with worries about new public uses of waterways.

From this point forward, the story of navigability in American law is too complex to trace in full. Navigability became important for a number of legal reasons, including the power of Congress to regulate commerce under the commerce clause, the jurisdiction of admiralty courts, and title to land. As each precise legal issue came up, the United States Supreme Court tended to talk about navigability in slightly different ways, adding further confusion to the whole legal field. Meanwhile, changes in America's economy were themselves raising new legal problems. In England, water travel and transport had rarely raised any concerns. In America of the nineteenth century, however, waterways were being used in new, more troubling ways. The right to transport was being called into question. Timber harvesters began floating logs to market, with the logs clogging waterways and often doing considerable damage along the way. Perhaps inevitably, disputes over log floating became framed in terms of a river's navigability, rather than, as they might have been, in terms of whether a particular waterway was a public highway, or whether the use was appropriate only if the user paid damages for the injuries caused to riparian landowners.

Mining companies aggravated this problem when they began dumping their rock tailings into waterways. Mining companies sometimes asserted that these, too, were uses of navigable waterways even when they clogged rivers so badly that they became nonnavigable in fact! Also producing litigation about navigable waterways were disputes over ice harvesting. Did the public's rights in waterways include the ability to harvest ice, or did the ice in a river belong instead to the landowners on either side? (An important case about ice reached the Illinois Supreme Court in 1881.[5] It involved ice on the Calumet River south of Chicago, right where George H. Hammond was setting up his meatpacking plants to start shipping iced beef to the East Coast. The court held that the ice belonged to landowners who held title to the underlying land, whether or not the river was navigable.)

Adding confusion to all of this was a series of rulings about waterways and land titles handed down by the United States Supreme Court, beginning with its 1842 ruling in *Martin v. Waddell's Lessee*[6] and its 1845 ruling in *Pollard's Lessee v. Hagan.*[7] These decisions established four points, as follows.

States Own Riverbeds

Because the king in England at the time of the Revolution owned all land beneath navigable (that is, tidal) waters, the original American states gained title to such lands from the moment of independence. The citizen-run states became the new sovereign powers.

All States Have Equal Ownership

Because all new states entering the Union thereafter entered on an "equal footing" with the original states, they, too, then, gained legal title to all land underneath navigable waterways the moment they entered the Union. This surprising, unprecedented legal conclusion produced an especially odd pattern of land ownership in early America, remaining to this day. When a state entered the Union, land on both sides of a river was often still in federal hands. The new state gained ownership only of the land beneath the river, a ribbon of state property surrounded by federal land. To complicate the matter still further, the state's title was subject to the paramount power of the federal government under the commerce clause to control the use of the river to maintain its utility for commercial navigation.

Navigability in Fact Prevails

Because so many American rivers were physically navigable well above the area of tidal influence, it only made sense for the nation to adopt a navigability-in-fact test, jettisoning the old tidal navigability test. And so the Supreme Court did. When states entered the Union, then, they immediately became owners of land beneath all waterways that were either subject to the tides or navigable in fact.

States Gain Title Subject to Public Rights

As for these various submerged lands, the states gained ownership to them under the "equal footing doctrine" in something like the same way that the king in England owned lands beneath navigable waterways there. That is, the land titles of states were subject to the public's rights to use the waterways. They were also subject to the severe constraints that had existed in England on the king's power to grant exclusive rights in the waterways to any private party. What the king could not do, the states apparently also could not do.

****

This line of legal reasoning no doubt looked good on paper. In application, though, it fostered serious troubles. Among the most obvious problems was that, by the time the Supreme Court got all these legal points straightened out late in the nineteenth century, most states already had transferred title to submerged lands to adjacent landowners. States were willing to protect the public's interests in such waterways, just as the Supreme Court said they should. But the states did not go back and revise land titles. They did not take back from riparian landowners their ownership rights in submerged waters. Nor did they change the land title rules prospectively. They continued their practices of granting title to submerged lands to adjacent landowners. Only western states, those just getting established when the Supreme Court's rulings were handed down, were in a position to embrace the Court's new legal reasoning. In the western half of the country, as a result, the predominant practice was for states to retain ownership of these submerged lands. When landowners in these states bought riverfront land they gained title only to the edge of the river, usually defined as the mean high-water mark.

By the late nineteenth century, most inland fisheries had become so degraded by pollution, siltation, and physical blockages that the economics of fishing had shifted. And with that shift came a predictable decrease in litigation involving fishing rights. As a result, courts handed down fewer legal rulings. Courts thus did little to clear up the law of public fishing rights. Navigability had always been about fishing rights, and navigability had now been expanded to cover inland waterways that were usable for travel. So did this mean that the public had fishing rights in all inland waterways? Apparently yes, but few courts addressed the issue directly. A related legal question: what limits did landowners face in their use of submerged lands when they obtained title to it? Did they face the same limits that the king did in England on the eve of the Revolution, or had the law changed? The United States Supreme Court provided no good answers. Surprisingly few state courts had much to say about the issues either. When courts did consider cases, the lawyers presenting them were often confused about the law, given its obscurity. Many judicial decisions considered only fragments of the complex legal regime that governed waterway uses. Courts handed down rulings on one narrow legal issue, not realizing that the disputes being resolved also implicated other legal issues.

RIGHTS TO FISH TODAY

Given this confusing legal history, where, then, do things stand today? What rights does the public have to fish, gather shellfish, and perhaps hunt waterfowl in inland rivers and lakes? The answers vary in their details among the fifty states, so much so that they cannot all be considered. In basic outline, however, the law is relatively coherent.

First, the law of each state includes a definition of navigability, usually termed something like "navigability for access." Any waterbody that is open to the public under this state law definition is available for public use. This is true without regard for the ownership of the land underneath the waterway, which varies among the states (as noted above, landowners in the East but not the West tend to own the land beneath navigable waterways). Most states expressly provide that the public's rights include the right to fish. In some states, this legal point is set forth by statute; in other states, in judicial rulings; and in a few states, in the state constitutions. The definitions of navigability that states use vary rather widely. Some states require a showing that a river can support commercial boat traffic for it to be navigable as a matter of state law. Some states continue to use the log-floating test developed in the nineteenth century. Still others define navigability in terms of whether a recreational boater can get a small watercraft down a stream. As for waterway obstructions, dams, and their effects on navigability, the common practice is to look to the condition a river was in before dams were built, and to ask whether the waterway could have been made navigable with reasonable improvements to it; if so, the river is navigable today, even if it is now blocked by dams and even if snags and other obstructions have not been removed. Some states, recognizing that recreational water use is itself now an important commercial activity, have expressly defined navigability in terms of recreational boat access, opening even very small creeks to public use.

A judicial ruling, typical of states with rather broad public use rights, is the Wisconsin Supreme Court's decision from 1949 in *Munninghoff v. Wisconsin Conservation Commission.*[8] In the case, a landowner sought a state permit to install muskrat traps on land he owned beneath a navigable waterway. The state commission denied the permit on the ground that it could not give out exclusive harvesting rights in a navigable waterway. Rights to use navigable waterways, the commission asserted, were reserved for the public. The state high court, hearing an appeal to the ruling, overturned the commission's decision. The court did so, not because of any flaw in the commission's reasoning, but because the public's rights in the waterway did not extend to the affixing of muskrat traps. "In general," the court ruled, "the rights of the public in the incidents of navigation are boating, bathing, fishing, hunting, and recreation." Trapping was not included, at least when traps were attached to the land. Trapping was, instead, "an incident of land use." Since the landowner held the rights to trap, not the public, it was permissible for the state commission to give the owner a permit for the traps.

Munninghoff illustrates the issues that have arisen in various states over the uses of waterways. Some states, such as Wisconsin, extend public rights to hunting as well as fishing. In other states, the right to hunt is either not included as a public right or, more often, not clearly established. In many states, fishers and (where allowed) hunters typically cannot affix structures or trotlines to private land. Only the landowner can attach things to the land. Public users, however, can portage around waterway obstacles and walk in the riverbed itself. In a few states, such as Louisiana, the public can make more extensive use of privately owned banks in connection with their waterway activities. Finally, in many western states, the state's ownership of the land itself extends to the mean high-water mark, which gives the public extensive rights to use land beneath that mark.

Although *Munninghoff* represents the dominant state law approach in the United States, allowing free public fishing in all waterways that are navigable in fact, a few states do seem to embrace still the old English common law rule in which public fishing is limited to tidal waters. New York's highest court in 1997 reaffirmed the rule,[9] even as it noted (and reaffirmed at length in 1998[10]) that all waterways that are navigable in fact are open to public travel and transportation. In some states, fishing issues have been so rarely litigated that the only relevant precedents are over a century old. Many nineteenth-century rulings display the influence of the older approach, limiting public use rights to tidal waters. The dates of these rulings make them less reliable than they would otherwise be.

If state law alone governed on this issue, the story might end here. But it does not. The story goes on, and becomes even more complex, because federal law also plays a role. And when federal law applies, it overrides any conflicting state law. The basic story is this: a waterway that is open under federal law must remain open, despite any contrary rulings or wishes of the state itself. That much is clear. What is somewhat less clear is whether all waterways open under federal law are open for public fishing, despite state efforts to vest fishing rights in the owners of the riverbeds. Decisions have been few, and almost nothing is known about whether federal law does or does not open up waterways to public hunting.

The Public Trust Doctrine

To approach the federal law aspects of public fishing rights, we can recall that states gained ownership of the land beneath their navigable waterways the very moment they entered the Union. The states took title to the land directly from the federal government so as to be on an equal footing with the original thirteen states. The original states gained their ownership rights as successor to the sovereign property rights of the king. And the king's rights, in turn, were subject to the powers of the public to make use of navigable waterways, including the right to fish.

The implication of all of this was sketched by the United States Supreme Court in an important 1892 ruling in *Illinois Central R.R. v. Illinois.*[11] When Illinois gained ownership of its submerged lands, it gained only a limited title, the Court announced, subject to public use rights. As the Court phrased it, the state's title to submerged lands was "a title held in trust for the people of the state, that they may enjoy the navigation of the waters, carry on commerce over them, and have liberty of fishing therein, freed from the obstruction or interference of private parties."

This 1892 ruling resurrected the "public trust doctrine" in American law, a doctrine that has had, ever since, an uneasy existence, largely because it has been difficult to figure out where the doctrine fits into American law. Is the public trust doctrine a federal law limit on what the state can do? Is it instead a type of state law, perhaps grounded implicitly in state constitutions, which imposes limits on what state legislatures can do? Is it merely some version of the common law, which states can override whenever they like merely by enacting a new statute? In practice, the Supreme Court has left it to the several states to decide for themselves how to carry out this special trust. At other times, the Court has stated that the public trust doctrine actually imposes duties on each state. In an 1855 ruling, for instance, the Court stated that a state has the "duty to preserve unimpaired those public uses for which the [submerged] soil is held."[12] In 1988, the Court noted in passing that states had the power "to recognize private right in [trust] lands as they see fit."[13] But then in a 1997 ruling, it asserted that submerged lands within a state are "infused with a public trust that the State itself is bound to respect."[14] The Court, in sum, has left matters muddled. The public trust doctrine does have force and does limit what states can do, but the states themselves have some flexibility in applying it.

The public trust doctrine deserves mention here because it provides a legal tool to challenge laws of states that appear to restrict public uses of waterways, including rights to fish. The public trust applies to all lands beneath waterways that were navigable at the time a state entered the Union. As two scholars summarized that law in 1967, navigability for this purpose was determined "by the natural and ordinary condition of the water at the time, not whether it could be made navigable by artificial improvements. However, the fact that rapids, rocks, or other obstructions make navigation difficult will not destroy title navigability so long as the waters were usable for a significant portion of the time."[15] The waters had to be usable by the "customary mode of trade and travel on water," but that could include waters "as little as three or four feet deep that are geographically located so that they have been, or can be used by canoes or rowboats for commercial trade and travel (fur traders' canoes). . . ." Land underneath such a waterway is subject to the public trust. If a state, contrary to the language of the *Illinois Central* ruling, does attempt to cut off public rights of access to such waters, or public rights to fish in them, its action could be challenged as a violation of the state's duty to preserve and protect public rights.

Navigation Servitude

More important than the public trust doctrine, in terms of providing clear federal protection for public rights, is the body of federal law known as the federal navigation servitude. This body of law, based on the Constitution (the interstate commerce clause, it appears), directly secures "the exercise of the public right of navigation" over all navigable waterways, as well as "the governmental control and regulation necessary to give effect to that right."[16] The navigation servitude is more than a federal power to regulate. It is a "dominant servitude" (that is, an overriding easement) upon the title of all such submerged lands subject to it.[17] As the Supreme Court put it in 1992, it is "a permanent easement that [is] a pre-existing limitation" upon the title of all such submerged lands, including lands held by private owners.[18] The easement, explained a federal appellate court the next year, "gives rise to the right of the public to use those waterways."[19]

Although the navigation servitude is well established in American law, it is curiously overlooked when lawyers think about public uses of waterways, particularly public rights to use waterways for recreational purposes—the public use issue that has drawn the most attention in recent decades. Any waterway subject to this federal servitude is plainly open to the public, state law notwithstanding. The ownership of the underlying land, as the Supreme Court has made clear, is irrelevant. As the Court explained the situation in a 1956 ruling, rejecting a landowners' claim for compensation under the just compensation clause of the Fifth Amendment:

> It is no answer to say that these private owners had interests in the water that were recognized by state law. We deal here with the federal domain, an area which Congress can completely pre-empt, leaving no vested private claims that constitute "private property" within the meaning of the Fifth Amendment.[20]

As for what waterways are covered by the federal navigation servitude, that issue is well settled in the sense that the governing legal test has been set forth clearly. In various of its navigation servitude rulings, the Supreme Court has cited, as the governing test of navigability, a ruling that it handed down in 1874 in *The Montello:*[21] a river is navigable if it is suitable for commerce conducted by "vessels of any kind that can float upon water," including those "propelled by animal power," and "no matter what mode of commerce may be conducted." Even more useful than this language was the Court's conclusion in the particular case, that the Fox River in Wisconsin—a shallow, meandering midwestern river, naturally full of snags—was navigable all the way to its source. This was true, the Court said, even though travel required dragging shallow, flat-bottomed boats over rocks and portaging waterfalls.

Under the navigation servitude, then, the public has access to virtually all waterways that are boatable, without regard for state law. (This legal conclusion, it is important to stress, is by no means well known, particularly by state officials.) How far this public access goes, in terms of recreational boating in small waters, remains to be seen. An instructive ruling was handed down in 1997 by the federal district court in Georgia.[22] Under Georgia law, a waterway is open to the public only if it is wide enough and deep enough to carry large cargo vessels. Nonetheless, according to the federal court, the public had access under the navigation servitude to a whitewater river in north Georgia as a matter of federal law. The particular river in dispute in the federal case contained rapids, rocks, and shifting currents, and its water flow was sufficient merely for kayaks to use it after rains. Still, the public had a federal law right to use it, notwithstanding the desires of landowners to keep them off and notwithstanding the vastly more narrow state definition.

Courts have paid less attention to the *fishing* rights that the public possesses under the navigation servitude. As noted, navigation for centuries was publicly important mostly to protect the public's fishing rights, not to protect rights of travel. Various federal court rulings over the past quarter century have either decided or assumed that waters subject to the federal navigation servitude are open to public fishing. The United States Supreme Court applied the same assumption in a key 1979 ruling.[23] A 2007 appellate court seemed to think otherwise, at least in the case of water that overflowed private land that was apparently above the mean high-water mark, but the ruling overlooked key precedents and misunderstood the purpose of the servitude.[24] On balance, it seems true that waterways that are navigable under *The Montello* standard of navigability are open to public fishing as a matter of federal law, without regard for state law.

Other Applicable Federal Laws

In addition to the public trust doctrine and the navigation servitude, other federal laws seem to guarantee public rights to fish in navigable waters. One such statute is the Rivers and Harbors Act of 1899,[25] which courts have assumed protects public fishing rights, again without regard for state law. States in the old Northwest Territory continue to be governed by a federal statute enacted in 1789, which established and protected public rights in navigable waters.[26] The statute remains in legal force even though the states in the former territory long ago entered the Union. The same language was used in federal statutes creating various other territories within the current United States, and it was included as well in the statutes admitting various states into the Union. All these federal statutes remain in effect, even though they are typically not contained in the *United States Code* and thus are difficult for practicing lawyers and judges to find.

In states such as Wisconsin and Louisiana, there is simply no need to resort to federal law to determine public rights. Public rights are fully protected under state law. The various federal laws become important only in states that attempt to limit public rights, including public rights to fish.

As for a federal right to engage in *hunting* in navigable waters, that issue has received essentially no judicial attention. Such rights, one might stress, were an integral part of public use rights in waterways navigable at law when the federal Constitution was drafted. It would be consistent with that historical legacy for the navigation servitude to safeguard this right as well. Times, though, have changed. Hunting is now a recreation rather than an essential element of survival. Perhaps one day we will learn from courts whether that change in circumstances produces a change in federal law.

A final issue, in terms of public rights to fish and hunt on waterbodies, has to do with lakes that are considered nonnavigable. The rights to use them are held by adjacent landowners, except perhaps in a few states such as Massachusetts. But what rights does each landowner around the lake have? If we look to land titles, the common practice is for landowners to own the land beneath the water to the center of the lake, to the extent it can be determined given contorted lake dimensions. Owners thus often possess odd, pie-shaped pieces of submerged land, tapering to a lake's center. The common law rule on such lakes was that each owner had exclusive control over the surface of the lake above that pie-shaped piece. Accordingly, no one had the legal right to use the entire lake surface without gaining the express or implied permission of all other landowners.

Disputes over nonnavigable lakes have rarely reached appellate courts. The law on the subject is therefore sparse. An indication of where the law might be heading on the issue was offered by the Illinois Supreme Court in 1988.[27] Because the case presented the first occasion ever for the Illinois court to address the specific legal issue, the court enjoyed greater freedom to fashion a rule that it believed made sense. After study, the court decided to ignore the English common law on nonnavigable lakes and to adopt instead the rule commonly followed on the European continent, the so-called "civil law rule." Under that rule, each landowner has the right to make use of the entire lake surface. That right is held in common with all other littoral landowners. Each landowner is limited to a "reasonable use" of the lake surface, consistent with the equal rights of other owners also to make reasonable use of it. In 1999, the Indiana Supreme Court faced a nearly identical legal dispute, involving a smaller nonnavigable lake.[28] That court decided to retain the common law and to reaffirm that individual landowners each owned a discrete portion of the lake. For landowners to have full use of the lake, they had to negotiate and come to some shared understanding.

On the subject of lakes, it should be emphasized that the navigability of a lake has nothing to do with whether the lake has a public boat ramp. Ramps are sometimes constructed on nonnavigable lakes, giving the public physical access to the lakes. Public uses of such lakes, however, could well violate the property rights of littoral landowners, who could sue for trespass or public nuisance. The reason that public use of a lake might violate the property rights of a lakefront landowner is as follows: When a government body buys a piece of land on a nonnavigable lake, it only buys the property rights possessed by the selling landowner. If the selling landowner only possesses rights to a wedge-shaped piece of the lake (as in Indiana), then the public only acquires that limited right. For the public to gain rights to use the entire lake surface, the government would need to purchase rights from all the lakefront owners, not merely the owner of the land where the ramp is installed. In Illinois and other states that follow the civil law rule, public users, on the other, would have access to the entire lake surface. But in those states, the public's use of the lake, considered as a whole, would still have to be reasonable in relation to the uses of the lake by other landowners on it. If many public boaters use the lake, crowding its surface, then the public's use could violate the property rights of other lake owners.

A final point: If a lake is navigable, the public, as we've seen, has rights to use the lake surface. To actually use the lake, however, public users need to have legal access to it, without trespassing. That issue itself, of public access, can prove difficult. In practice, the public can effectively lose its rights to use a particular lake if there is no valid public access to it. In such cases, government could exercise eminent domain to purchase a public right of access. Presumably the public can enter or exit a navigable water body wherever a public road touches upon it or goes over it. Roads, though, are usually constructed on road easements owned by government; the underlying land is still owned by adjacent landowners. The public's rights to use the road are limited to those activities within the legal scope of the easement that the government has purchased. Does a road easement include the right to move from road to river?

In all likelihood, a court that faced the issue would confirm that the public's use of a road includes the ancillary right to gain access to navigable waterways, to get off the road and put a canoe or rowboat into the river. Such a use of the road would seem consistent with the purpose of a road, which is to allow people to get from one place to another place. Over the past decade or two, disgruntled landowners around the country have raised this legal issue, claiming that public boaters have no legal right to use a road right-of-way to gain access to navigable rivers. If the issue leads to appellate litigation, we may soon have rulings that clarify the law.

LEGAL PROTECTIONS FOR FISHING RIGHTS

People engaged in fishing sometimes are physically disturbed or have their fish catches reduced by the misconduct of others. What legal protections do they enjoy and when can they recover damages for the lost fish?

Riparian Rights

The fishers who enjoy the clearest legal protection are those who own land along rivers and who fish from their private lands. Such riparian landowners possess riparian property rights to make use of these waters, including the right to fish in them, without regard for whether the waterway is navigable. These property rights exist in all states, even states in which the right to divert water is governed by the much-different prior appropriation system of water allocation.

Riparian landowners generally have the right to make reasonable use of their riparian rights, consistent with the rights of other riparian owners to use the waterway as well. A riparian landowner can therefore sue to halt another riparian who is using the waterway unreasonably, as by adding too much pollution. Late in the nineteenth century, courts ruled that a riparian cannot use a river in the way that would kill the fish in it. As the California Supreme Court put it in 1897, a riparian landowner "does not own the fish in the stream. His right of property attaches only to those he reduces to actual possession, and he cannot lawfully kill or obstruct the free passage of those not taken."[29]

Riparian water law gives landowners the power to complain about pollution that materially disturbs their use of a river, including pollution that harms wildlife. In a 1981 ruling, the West Virginia Supreme Court upheld the power of riparian landowners to challenge an upstream construction project that threatened to degrade the fishery.[30] A similar ruling was handed down by a federal court of appeals in 1975, in a case arising out of North Carolina. In that case, *Springer v. Joseph Schlitz Brewing Co.,* riparian landowners were allowed to sue an upstream brewery for pollution that "caused six unprecedented fish kills and otherwise impaired the quality" of the waterway.[31] The court explained the law as follows:

> In North Carolina, a riparian landowner has a right to the agricultural, recreational, and scenic use and enjoyment of the stream bordering his land, subject, however, to the rights of upstream riparian owners to make reasonable use of the water without excessively diminishing its quality. Though he does not own the fish in the stream, the riparian owner's rights include the opportunity to catch them. Interference with riparian rights is actionable tort, and a riparian landowner may join several polluters as joint tort-feasors.

Worth notice in this quotation is the court's reference to "scenic use and enjoyment" of the river by riparians, a right that other states routinely recognize as well. Thus riparians who use rivers to watch rather than catch wildlife have the same protections against interference with their wildlife activities.

Related to the issue of pollution and water quality is that of waterway obstructions, which can similarly interfere with a riparian landowner's rights to enjoy fish. Few cases have considered the issue in recent decades, although considerable legal precedent does exist from generations ago. An illustrative case is *State v. Haskell,* handed down by the Vermont Supreme Court in 1911.[32] The court explained the limited rights that riparian landowners have to degrade waterways:

> [A riparian landowner] cannot lawfully kill, materially injure, or obstruct the free passage of, those [fish] which he does not take. . . . He cannot divert [the waterway] from its course, nor pollute it, nor leave it

> so the landowners on the stream above and below may not enjoy a like use of the water, including taking fish therefore; and . . . this right carries with it the common right to have fish inhabit and spawn in the stream, for which purpose they must have a common passageway to and from their spawning and feeding grounds.

When a riparian landowner exceeds these limits on degrading a waterway, other riparian landowners harmed by the degradation can sue for damages and for an order halting it.

Public Fishing and Public Nuisance

Much fishing, of course, is done in public waters rather than from riparian land. In that situation, fishers have no private property rights at stake; they merely have their rights to fish as members of the public. Still, the fishers do have legal rights they can assert if they are harmed in their fishing. The chief avenue for legal relief is the action for public nuisance, which the fishers can bring because of the "special injury" they suffer in the form of their lost fish. When pollution involves a violation of the Clean Water Act,[33] a public nuisance action becomes far easier to prove because the statutory violation provides forceful evidence that the defendant's conduct is unreasonable. In addition, state statutes sometimes give fishers express rights to recover damages for unlawful pollution, to go along with the common law action for public nuisance.

A public nuisance action might also be brought against a landowner whose land uses have caused polluted runoff or have otherwise altered water flows in ways that harm fish catches. These cases, though, are harder for a fisher to win because the landowner's conduct is unlikely to violate any particular law. When conduct violates no specific statute, it becomes more difficult to prove that the defendant's conduct has been sufficiently unreasonable to warrant having to pay for the harm caused.

Harm from Water Withdrawals

A further way in which fishing can be harmed or undercut is from diversions of water by water users, either upstream or downstream, that lower water levels in ways that harm aquatic life. The law of water rights is itself quite complicated, and water rights are a form of secure private property. But holders of water rights are always subject to limits on the ways they can use their water. Nearly everywhere, the holders of water rights are obligated to use water in ways that are either "reasonable" or "beneficial," if not both. A fisher who has been harmed by water withdrawals would likely have sufficient injury to challenge a water diversion that is causing it. The legal claim would be that the water diversion fails to abide by the laws that water uses be reasonable or beneficial.

When fishing is threatened not by an ongoing water use, but by a proposed new one, fishers have even greater opportunities to intervene to halt the withdrawals. New water uses in most states are possible only if the user gets a water withdrawal permit. Permits are issued only when they serve the public interest as determined by a state water agency. Fishers therefore have the chance to object to a proposed permit on the ground that a new water withdrawal would conflict with the public good.

CHAPTER

6

The Constitutional Framework

Wildlife law is created by various levels of government within our federal system—from the national, or federal, level through state and local governments. With so many governing bodies, troubles can easily crop up. Different lawmaking bodies can pass laws that clash with one another. A law from one level of government might frustrate policies promoted by another. Plainly, housekeeping rules are called for, rules that explain what powers each government has and that resolve or avert direct and indirect conflicts. The rules that perform these functions are contained largely in the United States Constitution and in their state equivalents. The Constitution, in addition, imposes limits on what all governments can do, designed to protect individual rights and property entitlements. In combination, these legal provisions supply an overall structure for wildlife law.

This chapter surveys the powers of the various levels of government to control wildlife and human-wildlife interactions. It then looks briefly at constitutional limits that apply to all levels of government—rules protecting free speech, for instance, and rules insisting that all citizens enjoy the equal protections of the law. The chapter's third and longest section explores the most contentious set of housekeeping rules, the constitutional limits on when and how far state wildlife laws can discriminate between citizens of one state and citizens of other states. To what extent can a state reserve its wildlife resources for state residents? Can it ban imports or exports of wildlife and wildlife products? Can it impose higher license fees and lower harvest quotas on nonresident hunters and fishers? The chapter concludes with comments about state laws and federal lands. When do state laws apply on federal lands, and what power does the federal government have to nullify or displace state laws?

THE POWERS OF GOVERNMENTS

Sovereign power in our federal system is located chiefly at the state level. When the nation came together, the states, not the federal government, succeeded to the broad sovereign powers exercised in Britain by the king and Parliament. That power is often termed the "police power" and includes extensive authority to enact laws and regulations promoting public health, safety, and welfare. States are governed by state-level constitutions that constrain what they can do. In general, though, states have the broadest powers of all governments. In addition, as we've seen in chapter 2, they are viewed as owners of all wildlife as trustees for the people as a whole, with powers and perhaps duties to protect wildlife from harm.

Local governments are widely understood to be creatures of state law. That is, they are created by state law and possess only those powers given them by the state. In effect, states delegate part of their sovereign power to local governments. A local government, for instance, may or may not have power to control fishing within its boundaries; to ban possession of dangerous animals; to prohibit the sale of live animals or animal parts; to require game ranches to obtain licenses; or to regulate forestry and protect wildlife habitat.

As for the federal government, it possesses few, if any, inherent powers. Its powers come directly from the Constitution. At one time, courts interpreted the Constitution in ways that gave the federal government only modest roles to act within the nation's boundaries, except on federal territories and in navigable waterways. That arrangement has changed dramatically, particularly since the Depression era of the 1930s. Courts have interpreted the Constitution in ways that now let Congress and the president assert vastly broader authority. It remains true that the federal government is one of limited, enumerated powers, but so broad are the enumerated powers that few areas of governance escape federal reach. Today, limits on federal laws have more to do with political considerations and congressional disinterest than with any legal limit.

The Treaty Power

From its founding until well into the twentieth century, the federal government took little interest in wildlife harvesting, even when it had power to do so. States were left to take charge, on federal lands as well as elsewhere. The federal government's initial involvement with wildlife began around the turn of the twentieth century, when it stepped forward to help enforce state wildlife laws. Under the Lacey Act (1900, with many later revisions), violators of state laws became subject to federal punishment when they crossed state lines or otherwise engaged in interstate commerce.[1] Soon the federal government began expressing concern about decreasing populations of waterfowl. Migratory game species were on the decline. Bird conservation required coordinated action at the interstate level. After a false step or two, Congress in 1916 entered into a treaty with Great Britain, acting on behalf of Canada, to protect migrating flocks in North America. To implement the treaty, Congress enacted the Migratory Bird Treaty Act (1918), which imposed direct limits on hunting and other interferences with migratory waterfowl.[2] The validity of the act was immediately challenged. Did the federal government have the power to regulate the taking of waterfowl, particularly given the states' ownership of them? Could it enter into the treaty and then enact a statute implementing the treaty?

These questions made their way to the United States Supreme Court. In a 1920 ruling, *Missouri v. Holland,* the Court roundly upheld the power of Congress to act as it did.[3] Article II, section 2, of the Constitution expressly granted to the president, "by and with the Advice and Consent" of the Senate, the power to make treaties. Once the federal government entered into a treaty, the Court ruled, it had the power under the "necessary and proper" clause of Article I to enact statutes to carry out the treaty. So long as the treaty was valid, the implementing legislation was also valid. The power to enter into treaties, in short, itself enabled the federal government to regulate wildlife, independently of whether it had power under any other clause. The Supreme Court's ruling displayed a pragmatic recognition of the need for federal action to protect border-crossing flocks:

> Here a national interest of very nearly the first magnitude is involved. It can be protected only by national action in concert with that of another power. The subject matter [migratory birds] is only transitorily within the State and has no permanent habitat therein. But for the treaty and the statute there soon might be no birds for any power to deal with. We see nothing in the Constitution that compels the [federal] Government to sit by while a food supply is cut off and the protectors of our forests and our crops are destroyed.

The treaty power is one source of federal authority to protect wildlife, to the extent necessary to carry out duties pledged to other nations.

The Property Clause

An additional source of federal power comes from the clause in Article IV that gives Congress the power "to dispose of and make all needful Rules and Regulations respecting the Territory or other Property belonging to the United States." This power has proved especially important with respect to wildlife on the nearly 30 percent of the United States owned by the federal government—the national forests, national parks, wildlife refuges, and vast lands of the Bureau of Land Management. For decades after the nation was formed, many legal observers assumed that the federal government's extensive powers to control its territories largely disappeared when territorial land became a new state. Upon a state's entry into the union, the argument went, the federal government's power over land within the state diminished to that of a private landowner. At that point, the federal government was subject to state laws like all other landowners.

In a series of rulings beginning in the 1830s, the Supreme Court recognized, step-by-step, that the federal government had powers both as a landowner and as a sovereign. In an 1897 ruling, it allowed federal agencies to ban fences on private lands that effectively restricted public access to the federal lands.[4] In a 1927 ruling, it upheld a federal regulation restricting fires on private lands *near* national forests.[5] A bigger step came in 1928, when a dispute reached the Court over the management of the burgeoning deer herd in the Kaibab National Forest, just north of the Grand Canyon.[6] Arizona game laws severely restricted the taking of deer by anyone. Ignoring the state law, federal land managers proceeded to kill deer to thin the herd. The state arrested three of the managers. On review, the Supreme Court in *Hunt v. United States* summarily upheld the federal government's power to govern its land. "The power of the United States thus to protect its lands and property," the Court intoned, "does not admit of doubt . . . the game laws or any other statute of the state to the contrary notwithstanding."

The ruling in *Hunt* dealt with the ability of federal agents to protect federal lands against physical decline. Did federal power also extend to laws protecting wildlife itself, without regard for the land? The issue was more difficult because of the states' longstanding ownership claims in wildlife. Nonetheless, the Supreme Court had little trouble with the question when it reached the Court in *Kleppe v. New Mexico* (1976), to date its most important ruling on the scope of the property clause.[7] The federal law under review was the Wild Free-Roaming Horses and Burros Act of 1971, which protected "all unbranded and unclaimed horses and burros on public lands of the United States" from "capture, branding, harassment, or death."[8] New Mexico grazers who held grazing permits on federal land complained that wild burros were disrupting their livestock. State agents, responding to the complaints of the private grazers, entered federal land, rounded up the burros, and sold them at auction. The federal government argued, in its defense of the statute's validity, that the federal Horses and Burros Act indirectly protected federal lands; protection of the wild species, it asserted, helped "to achieve and maintain a thriving natural ecological balance on the public lands." The Supreme Court found it unnecessary to rely upon this land protection line of reasoning in order to uphold the statute's validity. To the surprise of many, the Court described the federal government's power as vast and extensive—indeed, almost without limits. The property clause, it explained, gave Congress authority to make its own determinations as to what were "needful" rules "respecting" federal lands, whether or not the rules had to do with the land's physical use or condition. "And while the furthest reaches of the power granted by the property clause have not yet been definitely resolved, we have repeatedly observed that '[t]he power over the public land thus entrusted to Congress is without limitations.'"

Since *Kleppe,* the federal government has encountered little resistance whenever it has asserted control over federal lands. Dissent over this broad federal power has continued in various western states, some of it labeled the "Sagebrush Rebellion." But federal law appears clear, and courts have routinely respected it. The chief lingering issue under the property clause has to do with Congress's power to regulate activities on nonfederal land to protect federal lands against physical harm and to halt nonfederal activities that might frustrate federal programs. The issue was posed by the Horses and Burros Act, which protected animals on private as well as public land, but the Supreme Court in *Kleppe* deemed it unnecessary to resolve the legal issue. Lower courts that have confronted the question since then have tended to uphold federal laws that regulate adjacent nonfederal lands. The federal government, though, has not exercised this power with any vigor; indeed, critics have accused it of doing far too little to protect federal lands against surrounding, inconsistent activities.

Litigation to date on this federal power has mostly dealt with federal laws that apply to adjacent *state-owned* land, not private land, particularly state-owned lands underlying navigable waterways. In one 1977 case, an appellate court upheld federal laws that banned hunting on lakes within a national park, even though the state owned the underlying land.[9] Another ruling in 1994 held that the federal government could prohibit the use of electronic fish finders on navigable waters that overlay state land located within a federally protected wilderness.[10] In 2004, the federal court in Colorado upheld the power of the U.S. Forest Service to require a special-use permit for a marina and boat docks constructed on private land immediately next to a national recreation area when the private lands were connected by channel to a federal lake.[11]

The Commerce Clause

The treaty and property clauses pale in significance as sources of federal power to the clause authorizing Congress to regulate foreign and interstate commerce. This power appears to extend to the regulation of all manner of human activities directly or indirectly linked to commercial activities that affect more than one state. Thus the federal government can regulate fishing if either the fish or fishers cross state lines or if the fish is sold in markets linked to interstate trade, as all markets are. The issue of Congress's power to regulate wildlife under the commerce clause didn't reach the U.S. Supreme Court until 1977.[12] The dispute dealt with coastal fishing. Virginia sought to limit the ability of nonresidents to fish off its shores. The Court held that Virginia's law was preempted by an inconsistent federal statute. To reach that conclusion, the Court had to uphold the validity of the federal statute that licensed fishing vessels. There was "no question," the Court concluded, that Congress could regulate fishing whenever there was some effect, even indirect, on interstate commerce. That power existed even if the fish themselves never crossed state lines. A 1979 federal court ruling upheld a federal statute dealing with airborne hunting.[13] The power of Congress to enact the law, the court announced, was not thwarted even when the statute banned activities that had no apparent connection to any interstate trade. Congress had the power to conclude that a class of activities affected interstate commerce "without the necessity of demonstrating that the particular transaction in question has an impact which is more than local."

In recent decades, only two types of wildlife-related disputes have raised serious questions about whether Congress has exceeded its commerce clause power. The Endangered Species Act,[14] as we'll see, prohibits actions that "harm" listed species. Harm includes activities on private land that degrade critical wildlife habitat so as actually to kill or injure members of a listed species. The regulation in effect restricts uses of private land. In several lawsuits, private landowners have challenged application of the federal regulation when the species itself has no commercial or trade value; when it exists only within the confines of a single state; *and* when citizens do not cross state lines to view the species (as they do, for instance, with rare bird species). In such instances, is there a connection to interstate commerce that would empower Congress to take action?

The other category of cases contesting federal power has to do with federal efforts to protect wetlands from filling and draining. When wetlands are adjacent to navigable waterways or clearly connected to them hydrologically, federal power is clear. But what about wetlands that are hydrologically isolated so that water in them does not drain or percolate into interstate waters? Is the interstate connection adequate if migratory birds use the wetlands? Is it enough if birdwatchers cross state lines to visit?

Neither of these issues has been the subject of a U.S. Supreme Court ruling. As for the Endangered Species Act, lower-court rulings have embraced a variety of arguments to uphold the federal regulation protecting rare species, even when a species exists only in a single state and lacks commercial value. As for the wetlands disputes, the Supreme Court has avoided the issue by interpreting the main federal wetlands statute—section 404 of the Clean Water Act[15]—in such a way as to curtail its application to isolated wetlands. According to the Court, Congress did not intend to apply the Clean Water Act to such lands. Should Congress expand the statute's coverage, clearly covering isolated wetlands, the Supreme Court will then need to decide whether it possesses adequate federal power to do so.

Congress has shown little interest in developing a comprehensive wildlife code. It has had adequate power to sustain the relatively few narrow statutes it has enacted. Questions of federal power could well loom larger if the federal role becomes more involved. Various conservation groups, for instance, have proposed creating vast interstate corridors of wildlife habitat to allow migration of species in response to climate changes. Presumably, planning for such corridors would require federal involvement, perhaps even extensive federal land use regulations. Should this happen, courts may have occasion to say more about the outer limits of Congress's power.

BROADLY APPLICABLE CONSTITUTIONAL LIMITS

The power of all governments in the United States to act is constrained by a variety of constitutional provisions aimed at structuring governmental power and protecting individual rights. These provisions apply in the wildlife arena just as they do to other areas of government action. Occasionally wildlife laws raise issues of free speech, protected under the First Amendment of the Constitution and under similar state provisions. We considered one such dispute in chapter 3, the cases challenging, as abridgements of free speech, state hunter harassment statutes that limit animal welfare activists in their efforts to dissuade hunters from killing animals. Occasionally, too, a dispute raises issues about freedom of religion. Thus an Amish hunter in 1996 challenged Ohio State's ability to require him to wear bright-colored clothing, claiming—unsuccessfully it turned out—a violation of his religious autonomy.[16] With the Supreme Court's recent decision that the Second Amendment right to bear arms is an individual right,[17] a rash of cases challenging game law limits on carrying loaded weapons during hunting season or on highways is foreseeable. State law equivalents of the Second Amendment also sometimes impose constraints.

Equal Protection of the Laws

Statutes operate by defining categories and applying legal consequences to those categories. When people are disadvantaged by being placed in a category, they may argue that they are being denied equal protection of the laws. Courts are particularly suspicious when they come across laws that distinguish among citizens based on classifications that are "suspect" or quasi-suspect. Thus laws that distinguish based on race or ethnicity and, to some extent, on sex—given the nation's history of discrimination on these grounds—are scrutinized to see if the state really had a strong enough interest to justify the distinction. Rarely, if ever, do cases involving such factors arise in the wildlife context. Instead, distinctions that appear in wildlife laws tend to turn on other differences among people: on their ownership of land in the state; on their residency in the state; or on where and when they want to harvest game. When laws distinguish among people on these bases, courts give them far more deference. Indeed, legislatures are rather free to draw lines among citizens, lands, and types of activities as they see fit, as long as they have a rational basis for doing so. It is not the court's job to second-guess what lawmakers do.

A wildlife law that did distinguish among citizens on a suspicious basis arose out of the emotional turmoil of World War II, reaching the Supreme Court in 1948.[18] A lawful resident of California of Japanese ancestry challenged a state statute that allowed only citizens to hold commercial fishing licenses. The Court struck down the statute as a denial of equal protection, given the state's inability to offer a substantial justification for the distinction. More typical of equal protection cases was a 2001 dispute resolved by the federal court in Montana.[19] A state statute, approved by ballot initiative, banned the hunting of game animals on game farms for a fee, while allowing other types of hunting. Game farm owners claimed that the law discriminated against them. The legal test used by the court in assessing the complaint was whether the statutory distinction—between game farms/fee hunting and fair-chase hunting—was "rationally related to further a legitimate state interest." To uphold the law the Court did not have to agree with the wisdom of it. It was enough if the state had a plausible policy reason for its classification, as it did in this instance.

In 2005, a federal appellate court upheld the validity of a Wyoming hunting statute that charged higher license fees for out-of-state hunters, reserved a disproportionate number of hunting licenses for in-state residents, and required out-of-state but not in-state hunters to hire guides to hunt in wilderness areas.[20] The statute was challenged by Donald Schutz, a longtime Wyoming resident who had moved to Florida. In rejecting his claim of unconstitutionality, the appellate court made clear how difficult it is to challenge a legislative classification that is not based on one of the few constitutionally suspect classifications:

> While we agree with Schutz that some of Wyoming's goals [in drawing the residence-based distinctions] may seem speculative, we cannot conclude that there exists *no* reasonable justification for the in-state preferences. Many reasons exist, in fact, for states to adopt a preference scheme. Residential preferences are commonly considered a benefit of state citizenship for finite resources such as wildlife resources, higher education, or access to state run facilities. While the reasons for preferences are varied—and context specific—it is not irrational to provide them. In-state residents, for example—especially those who hunt or fish—have a vested long-term interest in the sustainability of Wyoming's wildlife management system. This includes not just political support for such programs, but direct financial support through fees and taxes. In-state residents may be counted on more reliably to hunt in Wyoming year after year, thus supporting long-term game and fish habitat preservation, herd management programs, new species programs such as the introduction of the gray wolf or grizzly bear populations, or, finally the more mundane aspects of wildlife programs such as adequate highways, off-road and hiking trails, fire protection, and search and rescue programs.

In a 2006 ruling, another federal appellate court rejected an equal protection challenge to a federal regulation that gave to rural Alaskans a priority in the allocation of rights to hunt for subsistence purposes.[21] The regulatory aim was "to protect the subsistence way of life for Native and non-Native rural Alaskans." The court viewed the aim as legitimate and concluded that "a plausible policy reason" supported the classification.

The Just Compensation Clause

Another constitutional provision that limits wildlife lawmaking at all government levels is the clause in the Fifth Amendment of the Constitution that prohibits the federal government from taking private property from owners without paying "just compensation." For over a century, this restriction has also applied to state and local governments, by way of the Fourteenth

Amendment. The ban on takings was intended to apply to situations in which government physically seized property from an owner. In recent generations, courts have applied it also to laws and regulations that curtail the rights of property owners so severely that owners are essentially deprived of their property. This is the "regulatory takings" doctrine. It is the source of considerable political controversy, especially among critics who claim that the Supreme Court has given government too much power to invade private rights.

The regulatory takings doctrine confuses many observers because it places a limit on the powers of government to regulate or redefine property rights, yet property rights themselves arise out of law and are prescribed by law. Without law, no property rights as we know them would exist. Moreover, legislatures and courts have amended and updated laws for centuries, property laws very much included. Change in the rules of ownership are thus the norm, not some new development. The regulatory takings doctrine, it is clear, does not halt this kind of continuing legal evolution; it does not keep lawmakers from redefining the elements of property ownership. Instead, it puts limits on that lawmaking work. It curtails changes in property laws that seem to go too far in upsetting private expectations or that, in effect, single out one or a few property owners for burdens that other property owners do not share.

As we have seen in earlier chapters, property rights in land and wildlife largely originate under state law. States have great freedom in defining private rights as they see fit, particularly with respect to wildlife. States need not make wildlife available for private ownership, and by and large they have not. For instance, a person usually cannot gain property rights in the many nongame species. When state law does allow private ownership in some manner, whether by being first to capture the animal or otherwise, it can define the resulting property rights as it sees fit. Those who capture an animal, for instance, may or may not have the power to sell it. Their legal power to retain possession may only last during hunting season or for some other short period of time. They may also face anticruelty laws while the animal is alive. To put the point more broadly, an owner gains only those property rights that the law authorizes. The same rule applies in the case of land. The rights of landowners are rather carefully prescribed, with various natural resources sometimes included, sometimes not, in the bundle of landowner rights. The just compensation clause and the regulatory takings doctrine do not limit how governments *define* private property rights in advance. Instead, they curtail the ability of government to *redefine* private rights that are already in private hands.

The leading United States Supreme Court ruling dealing with regulatory takings and personal property (that is, property that does not involve land and items permanently affixed to land) arose out of a wildlife setting.[22] The dispute, mentioned in chapter 3, involved the application of the Bald and Golden Eagle Protection Act[23] and the Migratory Bird Treaty Act[24] to eagle feathers and claws. New federal regulations went into effect banning sales and trades of eagle parts already in private hands. Private parties could still retain possession and use eagle parts they already owned. They could also donate them to museums or charge money for people to view them. The most profitable activity, however—selling the parts—was halted. According to the Supreme Court in *Andrus v. Allard,* these new regulations were legitimate and did not amount to a taking of private property:

> The regulations here do not compel the surrender of the artifacts, and there is no physical invasion or restraint upon them. Rather, a significant restriction has been imposed on one means of disposing of the artifacts. But the denial of one traditional property right does not always amount to a taking. At least where an owner possesses a full "bundle" of property rights, the destruction of one "strand" of the bundle is not a taking, because the aggregate must be viewed in its entirety. . . . In this case, it is crucial that appellees retain the rights to possess and transport their property, and to donate or devise the protected birds.

Other judicial rulings have made clear that government has broad power to ban the possession of wild animals that pose a threat to public health or safety. Owners typically remain free for a brief time period to transport the animals for sale out of state, but this option can be foreclosed if surrounding states also ban possession. Courts have also been willing to let governments seize personal property used in committing a game law offense, at least so long as the value of the property seized is not so great as to seem unconscionable. Thus fishers convicted of violating fishing laws can have nets, gear, and even boats seized for their misdeeds.

Far more common than complaints that wildlife laws take *personal* property are the complaints that new laws interfere with private rights in *land,* particularly rights to develop land that is wildlife habitat. Not surprisingly, the case law on this issue is extensive, given the wide array of land use laws. The Supreme Court has long explained that governments possess considerable power to redefine the legal rights of current property owners and regulate their activities. Only the most extreme interferences with private rights run afoul of the ban on uncompensated takings. Thus governments can protect wildlife on private land, even when animals cause damage. They can ban development or intensive land uses that would disrupt populations and migrations.

The governing case law tends to divide regulatory takings challenges into three categories.

One category involves cases in which new laws authorize either government agents or the public generally to make a "permanent physical occupation" of the private land. Such laws violate the landowner's right to exclude, which the Court in recent decades has elevated to a prime landowner entitlement. An occupation need not be continuous to violate this constitutional rule, but it does have to be of indefinite duration. Thus a law that permits the public to walk up and down a private beach would violate this rule, even though public walkers are not continuously present on the private land. Courts have not extended this rule, though, to cover wild animals that enter private land. Thus, wildlife laws are valid even when they prohibit landowners from constructing fences or otherwise keeping animals away. Landowners, in short, have no constitutional right to exclude wildlife.

The second category of cases involves those in which a regulation bans all economically viable uses of a private land parcel. Such a law is invalid and requires compensation unless the law builds upon some background principle of property law that already limits what landowners can do. Thus a law is valid if it bans an activity that would qualify as a common law nuisance, even if it leaves land without economic value. Lawmakers are not limited, though, to banning only activities that courts generations ago would have deemed a nuisance. The concept of nuisance evolves. New activities can be deemed harmful and banned even though they were not considered nuisances long ago. Furthermore, it can be argued that the public trust in wildlife discussed in chapter 2 is itself a background principle of property law given the lengthy history of restraints on landowners to protect wildlife.

Few landowners have succeeded in gaining compensation under this second category of judicial precedent because few regulations have the effect of depriving land of literally all economic uses. A ban on hunting, for instance, would rarely, if ever, leave a landowner with no other economic options. As courts apply this "total takings" rule, they look at the value of a land parcel as a whole. They do not consider how a regulation affects a particular portion of the land. Thus, a law that bans all development in a wetland would not require compensation if the wetland acres are part of a larger private tract that also includes good development land.

Most disputes about alleged regulatory takings do not fall in either of these two categories, either the "permanent physical occupation" category or the "total-takings" category. They fall instead into the third category and are resolved using a more flexible, amorphous multifactor test. The applicable test seeks to determine whether, in all justice and fairness, the economic burden of a particular land use rule should be borne by taxpayers rather than affected landowners. The multifactor test—first articulated in 1978 in a prominent ruling involving Grand Central Station in New York—instructs courts to balance the public and private interests at stake, paying particular attention to the economic impact of the regulation on the private owner, the regulation's interference with the owner's "distinct investment-backed expectations," and the "character" of the government action.[25] Courts applying this test have rarely concluded that a land use rule crosses the constitutional line so as to require government to pay compensation.

Occasionally, takings case have arisen that involve government permits and licenses. Generally, holders of permits and licenses have no rights to renew them unless the law expressly provides otherwise. In addition, if a license is not renewed or is terminated in accordance with its provisions, the licensee typically cannot recover consequential losses, such as the market value of fishing gear or a boat dock that has declined sharply due to license revocation. A typical case was handed down by the Court of Federal Claims in 2004, when it rejected a request by a fishing corporation for compensation.[26] The corporation claimed that a compensable taking occurred when Congress enacted the American Fisheries Act in 1998,[27] terminating the plaintiff's license to fish. The court concluded that the corporation had no protected property interest in its license and the government could thus alter or terminate it in the public interest. Further, the law did not effect a taking of the corporation's fishing vessel simply because the corporation couldn't use the boat in a particular fishery. Although the boat declined substantially in market value, the owner could still sell it or use it elsewhere.

FEDERAL LIMITS ON STATE POWER

The most-litigated constitutional disputes dealing with wildlife have been those calling into question the ability of states to write wildlife laws to favor their own residents. As we've seen, laws that distinguish between residents and nonresidents are subject to challenge under the equal protection clause. States are subject to additional restrictions by virtue of their membership in a union of states. Beyond that, they are subject to limits that arise because states are subordinate to the federal government whenever the federal governments acts pursuant to one of its enumerated powers. Laws promulgated at the federal level, accordingly, take precedence when they collide with state measures.

During the nineteenth century, the prevailing understanding was that states had especially strong power over wildlife, based on their ownership of it. That ownership, it was said, largely insulated states from federal interference in the ways the states went about allocating wildlife to citizens and defining the private rights hunters and fishers obtained. The state ownership doctrine, in short, provided a shield for states to resist federal laws that interfered with their wildlife operations. Over the past century, that shield has fallen away. The state ownership doctrine remains alive, as we saw in chapter 2, but it no longer insulates states from having to comply with the general structures of federal governance in the wildlife area, just as they do in all other areas of operations. That is, in the application of basic constitutional provisions, wildlife law is no longer different except insofar as courts take into account the lengthy history of state leadership on wildlife issues and the continuing tendency of Congress to let states take the lead.

Preemption

The federal government is supreme within our federal system. When valid, federal laws and regulations take priority over contrary laws adopted by lower levels of government. This rule, simple enough to express, has proved difficult to implement because conflicts are not always easy to identify. Hardly a year goes by without the Supreme Court's having to resolve a preemption case dealing with some area of law.

The body of case law that has developed in this much-litigated area instructs courts to consider two issues as they go about deciding whether a state law has or has not been preempted by a superior act of the federal government. The first issue has to do with legislative intent. When it enacted a particular law, did Congress intend to displace state law on the same issue? If it did, then the federal law displaces the state or local law. Congress might express its intent to preempt in a clear and direct way, right in a statute, or it might do so implicitly. As courts go about discerning the intent of Congress, they tend to look at a variety of factors. Did Congress seem to legislate in a way that was meant to be exclusive, leaving no room for states to get involved? Did Congress somehow imply a settled judgment on an issue so as to displace any role for state supplementation? When these questions arise, courts consider an array of evidence. They look at the language of the federal statutes, congressional hearings, and the sheer complexity and detail of the federal legislation itself. Ultimately it is a judgment call.

The second issue in preemption cases has to do with conflict, and whether the state laws being challenged collide with the federal law in some way. In a few rare instances, the conflict is so great that a citizen literally cannot comply with both laws at once, given their contrary guidance. More often, the conflict is less direct. It comes about because compliance with the state law would seem to undercut achievement of the purposes or goals of the federal legislation. This prong of the preemption test is often termed the "frustration of federal purposes" test because it looks to whether state laws, if enforced, would frustrate the desires of Congress. The difficulty that arises when applying this frustration-of-purpose standard comes at the first step, identifying Congress's purpose. To use the facts of a prominent ruling, when Congress prescribed the environmental standards to govern mining on federal lands, what was its intended purpose? Did it intend merely to ensure that any mining taking place was performed in an environmentally reasonable manner? Did it instead want to promote mining affirmatively while keeping its environmental costs within reasonable limits? If the intent of Congress was the former, then a state could add additional environmental protections without disrupting Congress's intent. If instead the intent of Congress was to promote mining that it deemed environmentally sound, then additional state protections might clash with the purpose, particularly if the additional protections made mining infeasible.

Because Congress has never enacted a comprehensive wildlife scheme, few state wildlife laws have been struck down on preemption grounds. Several categories of federal lands are closed to hunting, a policy that does preempt state laws that would otherwise allow it. In addition, federal land managers can undertake actions that would violate state fish and game laws if performed by a private party. Otherwise, federal-state conflicts, and thus preemption, are relatively rare. In a 2002 ruling, a federal appellate court concluded that statutes governing national wildlife refuges did preempt state management of wildlife on the refuges when state law conflicted with federal plans.[28] In the ruling, the court held that Wyoming was foreclosed from vaccinating elk on the National Elk Range with brucellosis vaccine when the vaccination conflicted with federal management guidelines. Also illustrative is a 2001 ruling from the federal court in Hawaii, concluding that a state law prohibiting vessels from navigating in certain waters was preempted by the conflicting terms of federal licenses that authorized the navigation.[29]

Interstate Discrimination, Direct and Indirect

The power of states to write wildlife laws as they see fit is also constrained by two specific clauses of the Constitution. The clauses were designed to unify the nation and keep states from fragmenting into individual, selfishly governed enclaves. One clause, the privileges and immunities provision of Article IV, announces that "the Citizens of each State shall be entitled to all Privileges and Immunities of Citizens in the several States." The other, the commerce clause of Article I, grants express power to Congress to regulate interstate commerce, leaving states free to regulate as well but implicitly prohibiting discrimination based on state borders.

We can see how these two provisions operate by considering the three types of state laws that occasionally run afoul of them: laws that ban possession and sale of wildlife and wildlife products; laws that ban their exportation from or importation into a state; and laws that discriminate between state residents and outsiders in the rules governing access to wildlife resources in the state, including access to hunting licenses, quota allocations, and hunting grounds.

On the issue of bans on possessing and selling wildlife and wildlife products, we have already seen that states possess broad powers to define property rights as they see fit. They can ban all sales and limit possession of wildlife to specified seasons. So long as such laws do not distinguish between state residents and outsiders but instead ban possession and sale by everyone, they do not overtly discriminate against interstate commerce. They are thus not subject to the special close scrutiny that courts apply when considering the validity of openly discriminatory laws. For instance, a federal appellate court in 1994 had no trouble upholding the validity of Washington State wildlife regulations that banned the private ownership and exchange of various species of exotic animals, including mouflon sheep and fallow and sika deer.[30] In resolving the case, the court summarized the legal test that it used to decide whether state laws violate the prohibition against discrimination implicit (or "dormant") within the commerce clause:

> If the regulations discriminate in favor of in-state interests, the state has the burden of establishing that a legitimate state interest unrelated to economic protectionism is served by the regulations that could not be served as well by less discriminatory alternatives. . . . In contrast, if the regulations apply evenhandedly to the in-state and out-of-state interests, the party challenging the regulations must establish that the incidental burdens on interstate and foreign commerce are clearly excessive in relation to the putative local benefits.

Given that the Washington ban on possessing and selling the exotic species did not overtly discriminate between in-state and out-of-state residents, it was subject to review under the deferential standard of the second part of the dormant commerce clause test. On the facts of the case, Washington's motive in promulgating the regulations was to protect native wildlife from diseases and parasites in order to maintain the genetic purity of wild species. Given these local benefits, the incidental burdens on interstate commerce were not clearly excessive. The regulations were thus valid.

The outcome of the Washington dispute is typical of cases involving complete bans on possessing and selling specific wild species. A different, harder case is posed when a state law bans only the sale of animal parts, including meat intended for sale. In a 2003 dispute, a seafood distributor challenged a South Carolina statute banning the sale of white bass within the state.[31] The ban applied to state residents as well as outsiders, and thus did not discriminate overtly. Like the Washington wildlife regulations, then, the South Carolina law was subject to the more deferential standard, which upheld laws so long as the burdens on interstate commerce were not clearly excessive in comparison with the local benefits. Because the white bass were not imported alive, they did not pose the same ecological dangers as did the wildlife in Washington. In addition, South Carolina had a similar statute that banned the sale of white perch within the state. This law, though, allowed seafood importers to bring perch into the state and sell it so long as they kept accurate records to ensure that no white perch caught within the state was also sold. A similar arrangement, it seemed, would have worked with respect to the white bass. Given the availability of this option, which would have eliminated the burdens on interstate commerce, the South Carolina statute was struck down as inconsistent with the commerce clause.

While complete bans on selling wildlife are almost always upheld, the opposite outcome is nearly universal when a state tries to allow sales of wildlife *within* the state while prohibiting sales *out* of state or sales intended for export. Such laws discriminate overtly based on state boundaries. They are constitutional only if a state can show a motive for them, unrelated to protectionism, that cannot be served by any less-discriminatory means. The important 1896 Supreme Court ruling in *Geer v. Connecticut* upheld a statute of just this type on the ground that the state as owners of the wildlife could define property rights in captured animals as it saw fit, including a definition of property rights that allowed sale within the state but not outside it.[32] As we saw in chapter 2, however, *Geer* was expressly overruled by the Supreme Court in *Hughes v. Oklahoma* (1979).[33] *Hughes* ruled that state ownership was an insufficient ground to immunize a state law from dormant commerce clause scrutiny. The state statute challenged in *Hughes* appeared to be driven by economic protectionism, and the state had no valid purpose that could not be met as well by less-discriminatory measures. The ban on out-of-state sales was thus invalid.

The ability of states to ban the *importation* of wildlife is quite similar to the rules governing bans on possession and sale. Complete bans on importation of wildlife are valid so long as the state has a legitimate reason to ban imports and so long as it also bans possession by state residents. When that is done, out-of-state sellers are not disadvantaged. The problem arises, again, when states seek to use their boundaries as a basis for distinguishing between permissible and impermissible conduct. Generally, simple bans on importation are void under the dormant commerce clause. Rarely can states articulate a reason why residents can buy and possess wildlife within the state but cannot import the same species from out of state. A rare, prominent exception to this legal outcome came in a 1986 Supreme Court ruling, *Maine v. Taylor*.[34] Maine allowed state residents to raise and sell baitfish but refused to allow live baitfish to enter from out of state. Its announced reason was a worry about accidentally importing fish parasites and fears that nonnative baitfish might be inadvertently included in the shipments. The nonnative fish could escape and disturb aquatic ecosystems in the state, competing with native fish for food or habitat and preying on native species. The state claimed that it was infeasible to test imported shipments for parasites and nonnative fish due to the small size of the fish and the large quantities involved. The Supreme Court accepted the argument and upheld the state statute despite the overt interstate discrimination. The state had shown a valid, substantial local interest, and shown also that no less-discriminatory means were available to guard against the dangers. In upholding the statute, the Court employed a type of precautionary rule to guard against environmental harms. The state, it explained, had "a legitimate interest in guarding against imperfectly understood environmental risks, despite the possibility that they may ultimately prove to be negligible."

When states use state lines as bases for their rules governing *access* to wildlife—as they often do—the legal outcomes tend to be more mixed than when they ban possession, sale, export, or import. The nineteenth-century approach, again, was to allow states nearly unlimited freedom in setting rules about hunting and fishing, given the state's ownership of the wildlife. The people of the state owned the wildlife collectively. Laws that limited use of wildlife to state residents merely regulated "the use by the people of their common property," as the Supreme Court explained in 1876.[35] Today, courts review state laws with care to ensure that they don't violate either the privileges and immunities clause or the dormant commerce clause. It remains true, however, that states retain unusual flexibility when they themselves undertake to provide benefits to citizens. States can provide free education, subsidies, and an array of other economic benefits to their residents while withholding the benefits from nonresidents. Presumably a state could distribute wildlife in somewhat the same way if it did so directly through public programs.

The limits imposed on states by the privileges and immunities clause have become relatively clear in recent decades. The Supreme Court explained the content of the clause in a prominent 1948 ruling, *Toomer v. Witsell,* a dispute involving a complex South Carolina regulatory scheme governing access to off-coast shrimp fisheries.[36]

> Like many other constitutional provisions, the privileges and immunities clause is not an absolute. It does bar discrimination against citizens of other States where there is no substantial reason for the discrimination beyond the mere fact that they are citizens of other States. But it

> does not preclude disparity of treatment in the many situations where there are perfectly valid independent reasons for it.

In *Toomer,* a South Carolina law required out-of-state shrimp fishers to pay a license fee one hundred times greater than the in-state fee. They also had to unload their catch for processing at a South Carolina port before removing it from the state. So discriminatory was the scheme that it essentially excluded out-of-state shrimpers from working South Carolina waters. The law was not necessary to curtail harvesting; other methods could be used to do that. The law, accordingly, was void under the privileges and immunities clause. In a separate concurring opinion, Justice Felix Frankfurter distinguished South Carolina's law from a law that provided a benefit to state residents from the state's natural bounty:

> It is one thing to say that a food supply that may be reduced to control by a State for feeding to its own people should be only locally consumed. The State has that power and the Privileges-and-Immunities Clause is no restriction upon its exercise. It is a wholly different thing for the State to provide that only its citizens shall be engaged in commerce among the States, even though based on a locally available food supply. That is not the exercise of the basic right of a State to feed and maintain and give enjoyment to its own people.

In 2003, a federal appellate court in *Connecticut v. Crotty* followed the guidance of *Toomer,* striking down a similar state licensing statute that overtly discriminated against out-of-state commercial lobster fishers.[37]

The strong constraints that the privileges and immunities clause imposes on a state do not apply to laws and regulations governing *recreational* hunting and fishing rather than *commercial* operations. The clause, the Supreme Court has explained, only protects nonresidents with respect to "privileges" and "immunities" "bearing upon the vitality of the Nation as a single entity." Put otherwise, the clause only comes into play when a nonresident complains about interference with some type of fundamental right such as the right to a person's livelihood. In *Baldwin v. Fish & Game Commission,* the Supreme Court considered whether recreational hunting was sufficiently fundamental to trigger application of the clause.[38] The dispute involved a challenge to a Montana hunting scheme that charged nonresidents license fees to hunt elk that were between seven and a half and thirty-seven times the cost of licenses to residents. The Court determined that the clause did not apply to recreational hunting:

> [The nonresident hunters'] interest in sharing this limited resource [elk] on more equal terms with Montana residents simply does not fall within the purview of the Privileges and Immunities Clause. Equality in access to Montana elk is not basic to the maintenance or well-being of the Union. [The hunters] do not—and cannot—contend that they are deprived of a means of a livelihood by the system or of access to any part of the State to which they may seek to travel.

The ruling in *Baldwin* created what is now a rather clear legal distinction between the regulation of commercial harvesting and recreational harvesting. The privileges and immunities clause only restricts the former. Even then the ban on discrimination is not complete. A state can defend even overt discrimination against nonresident commercial harvesters if it shows that the restriction is closely related to the advancement of a substantial state interest. In practice, this means that states can charge higher fees to out-of-state commercial licensees if they can show that the higher fees are set so as to allow the state "to recover from nonresidents their share of the expenses of managing the resources from which they are benefitting." This latter test was applied in a 1993 ruling evaluating a Virginia statute that set higher commercial licensing fees for nonresidents.[39] The court struck down the arrangement, not because Virginia was unable to charge more to out-of-state fishers, but because it had failed to "create any credible method for allocating the costs as between residents and nonresidents which places the burden equally or approximately equally upon residents and nonresidents."

Recent rulings have respected this distinction between commercial and recreational hunting and fishing, giving states considerable freedom to enact laws for recreational hunting as they see fit with little or no worry about the privileges and immunities clause. A 2006 ruling by the federal court in Kansas upheld that state's scheme although it sharply discriminated against nonresident hunters.[40] The complaining hunter, a New Mexico resident, owned considerable land in Kansas, paid taxes on it, and managed the land chiefly as a wildlife haven and hunting preserve. Kansas allowed resident landowners but not nonresidents to obtain "hunt-on-your-own land" permits. It allowed residents to hunt mule deer with rifles but limited nonresidents to muzzle-loaders. Turkey hunting permits, few in number, were reserved entirely for residents. Despite the plaintiff's ownership of land in Kansas, and even though the land was in an area designated for wild turkey hunting, the plaintiff was treated as a nonresident. The privileges and immunities clause did not invalidate the discriminatory state scheme since only recreational hunting was at issue. The plaintiff landowner tried to get around the recreational-commercial distinction by arguing that he deserved more favorable treatment as a landowner. The court, though, rejected that claim because landowners in Kansas, like landowners elsewhere, had no property right to hunt. The Kansas law therefore did not discriminate against the plaintiff in his ownership of land, merely in his recreational hunting on the land.

A federal appellate court a few months later handed down a similar ruling in a case in which Minnesota residents challenged a North Dakota hunting scheme that sharply favored local residents in allocating rights to hunt waterfowl.[41] North Dakota excluded nonresidents from hunting during the opening week of waterfowl season, charged higher license fees, and restricted hunting on certain public lands. Like the Kansas scheme, it exempted landowners from needing licenses to hunt on their lands during seasons, but only if the landowners were state residents. The court upheld the North Dakota law. The privileges and immunities clause did not apply because the hunting was purely recreational. Plaintiffs were not discriminated against in their capacities as state landowners because they did not possess any property right in hunting.

State laws governing hunting and fishing are also subject to challenge under the dormant commerce clause, just as are laws that deal with possession, importation, and export. The applicable legal test, as we have seen, contains two parts. If a statute overtly discriminates against interstate commerce, it is presumptively invalid. A state can defend it only by showing that the statute promotes a substantial state interest that cannot be served by a less discriminatory means. If the instead statute is facially neutral and the harm on interstate commerce only indirect, then the state has more discretion; it can defend a law by showing simply that the ill effects on interstate commerce are not "wholly disproportionate" to the local benefits.

State laws dealing with commercial fishing obviously relate to commerce. When these laws distinguish between in-state and out-of-state fishers, the laws are subject to close scrutiny and are likely invalid. States, that is, cannot reserve their resources for residents once they decide to open up the resources for public commercial access. It is not enough to point to a need to conserve the resources since conservation can be promoted with harvesting limits that affect everyone equally. States fare better defending their laws when the laws do not overtly discriminate but instead affect interstate commerce only indirectly. Thus in a 2001 ruling, the federal court in Montana upheld a state statute, adopted by ballot initiative, that banned new game ranches and restricted the shooting of ranch-raised game.[42] Because the statute applied to residents and nonresidents alike, it was not subject to the strict judicial review reserved for overtly discriminatory provisions. Instead, it was reviewed under the more deferential rational basis standard. The court ultimately upheld the statute, noting that its ill effects on interstate commerce were slight and that the state had made a convincing case that the statute was needed to protect against possible harms to state wildlife populations.

Less clear is the application of the dormant commerce clause to state laws placing nonresidents at a disadvantage when it comes to recreational hunting and fishing. Do recreational activities amount to interstate commerce for purposes of the dormant commerce clause analysis? The issue, we might assume, would be clear by now, but precedent on the issue is sparse. A prominent 2002 ruling by a federal appellate court concluded that the clause did apply.[43] The court remanded the dispute to the district court to make factual findings about whether the Arizona statute being challenged was adequately justified by state interests. The trial court was also to figure out whether less-discriminatory provisions could fulfill the state's purposes. Not much later, another federal court disagreed with this reasoning.

Uncertainty about the issue and disagreement with the 2002 appellate ruling prompted Congress to get involved. Congress added a rider onto a 2005 appropriations statute, the "Emergency Supplemental Appropriations Act for Defense, the Global War on Terror, and Tsunami Relief, 2005." The added section was termed the "Reaffirmation of State Regulation of Resident and Nonresident Hunting and Fishing Act of 2005." The brief statutory provision declared, as a matter of public policy, that Congress wanted states to exercise freedom in regulating hunting and fishing. It expressed approval of "laws or regulations that differentiate between residents and nonresidents of such State with respect to the availability of licenses or permits for taking of particular species of fish or wildlife, the kind and numbers of fish and wildlife that may be taken, or the fees charged in connection with issuance of licenses or permits for hunting or fishing." The statute provided that courts thereafter should not construe the commerce clause to restrict the state's ability to discriminate.

The dormant commerce clause is a restriction on the states, not on Congress. It has long been the law that Congress can authorize the states to do what otherwise would run afoul of the implied constitutional goal of creating a single, national market. Since the 2005 statute became law, two federal appellate courts have considered its legal effect. Both have interpreted it to remove the dormant commerce clause as a limit on the ability of states to legislate on hunting and fishing. States laws still might be preempted by federal law, or run afoul of the privileges and immunities or equal protection clauses. But the dormant commerce clause, it appears, is not at the moment a barrier to discrimination. Some uncertainty remains as to whether Congress intended the 2005 act to become permanent law or whether it was intended to endure only for the one-year period covered by the appropriations act. The statute on its fact has no time limit, however, and would seem to have continuing force, even though Congress did not give instructions to include the provision in the *United States Code.*

The end point of this discussion would appear to be as follows: States are free to draft *recreational* hunting and fishing laws as they see fit. They can distinguish between residents and nonresidents so long as their laws are supported by enough rational basis to survive equal protection scrutiny and so long as they are not preempted by a federal law on the subject. Neither the privileges and immunities clause nor (given the new statute) the dormant commerce clause imposes additional limits on discrimination. In the case of *commercial* hunting and fishing, states must comply with the privileges and immunities clause. What they need to demonstrate to do that remains rather unclear. Litigation so far has typically dealt with higher license fees, which can be charged to nonresidents only if the fees are reasonably designed to offset higher costs that the state incurs when nonresidents come to the state to hunt and fish. As for other legal restrictions, they must be justified by reference to their ability to serve state needs, though the exact legal standard remains uncertain.

WILDLIFE ON FEDERAL LANDS

A final issue to take up in this survey of constitutional structures and limits is the legal status of wildlife on federal lands. The federal government, as we've seen, has vast if not unlimited powers to legislate with respect to them.

What, though, about states and even local governments? Can they apply their laws to people and activities on federal lands?

It is useful to begin with a distinction between single-use and multiple-use lands. State laws are least likely to apply to single-use lands—most notably military lands. On the other hand, state laws are most likely to apply to multiple-use lands such as the national forests and the lands managed by the Bureau of Land Management. Between these extremes are dominant-use lands such as national parks and national wildlife refuges.

It comes as a surprise to many to learn that state and local laws generally do apply on federal multiple-use lands. A contrary result would cause all manner of legal confusion. Murder is not generally a federal crime, nor is there any federal law dealing with automobile accidents or product liability. From an early point, the federal government stood back and let states regulate the taking of fish and wildlife on federal lands. The federal government could have, but did not, adopt its own fish and game regulatory system. Federal agencies, though, have become involved in the issue in recent decades, as has Congress. Hunting is largely banned in national parks outside Alaska. Federal land plans in many places also put limits on when and where people can hunt and fish. The foundational legal principle here is that Congress and federal agencies can displace state and local laws when they choose to do so. In the absence of action, however, federal lands are treated the same as privately owned lands.

The legal rule that determines whether state laws apply on federal lands is the preemption doctrine, considered above. It applies in this factual setting pretty much the same way that it does in all legal settings. Courts, as noted, decide whether state law is preempted by looking at the intent of Congress to preempt, if any, and at evidence of conflict between the federal and state regulatory schemes. The leading ruling on the subject as applied to federal lands is a Supreme Court opinion from 1987, *California Coastal Commission v. Granite Rock Co.*[44] The dispute involved a mining operation on federal land, conducted in accordance with environmental guidelines set down by the U.S. Forest Service. California insisted that the mining company also obtain a permit and comply with regulations from the state designed to protect its coastal zone. Thus the issue: did the state laws apply, or were they preempted by the federal land management scheme with its own environmental protection provisions? The Supreme Court ruled that state law continued to apply. There was no congressional intent to preempt state action, the Court concluded, and state law did not frustrate any federal purpose. In reaching this conclusion, however, the Court assumed that a state could *regulate* mining but not *prohibit* it, and could impose *land use* regulations, but not *environmental* ones.

States, in sum, have broad powers to apply their wildlife laws on federal lands. They are subject to being overridden by Congress and by the land use plans of federal agencies. But unless Congress or federal agencies act, the full range of state laws applies. In all likelihood, states could designate specific federal lands as no-hunting zones. They cannot, though, tell federal agencies how to manage their lands, nor can they, of course, authorize hunting or fishing when federal law prohibits it.

CHAPTER

7

State Game Laws and Liability for Harm

English-speaking settlers had hardly arrived in North America when colonial governments began enacting laws that controlled how people hunted and fished. Shortages of deer appeared early, as did disputes over access to fishing and shellfish grounds. Even when shortages were not acute, it made sense, many colonists realized, to avoid disturbing wildlife during breeding seasons.

Predictably, early colonial laws built upon the game law traditions of the dominant home country, England. That tradition had begun many centuries earlier, before Parliament even existed. The king and his officers protected valuable wildlife directly, based on the king's claimed ownership. The king as owner set the rules on who could take what and when. Parliament began legislating as it gained power. A statute from 1285, for instance, restricted the taking of salmon during certain seasons. Repeated offenders were subject to imprisonment—"for the third Trespass, they shall be Imprisoned a whole year; and as their Trespass increaseth, so shall their Punishment."[1] Not later than the fifteenth century, Parliament began limiting the types of equipment that could be used to catch fish. Bag limits were also imposed, along with prohibitions on various types of commerce in wildlife. By the seventeenth century, laws restricted hunting to landowners who possessed significant wealth. A 1671 law limited hunting to those whose land yielded an annual income of at least one hundred pounds (when the annual wage was no more than four pounds).[2] The law authorized larger landowners to enforce this limit directly through their private gamekeepers, who were granted power to take from all violators their "guns, bows, greyhounds, setting-dogs, lurchers, or other dogs, . . . ferrets, tramels, lowbels, hays, or other nets, hare-pipes, snares, or other engines." The most notorious and hated act came in 1723, the Waltham Black Act, which imposed capital offenses for over two hundred crimes, including poaching on private land.[3] Although the Black Act was widely reviled in the colonies, colonies and early state governments continued to regulate hunting and fishing. They did so with various goals. Conservation was one, but there were various others, related to hunting ethics, the maintenance of peace, and promotion of the common good.

The modern regulatory regime, with its detailed rules, has emerged from this long and checkered history. Today the applicable regulations in many states comprise literally hundreds of pages, with detailed rules that vary by species and location within the state. State laws and regulations also differ widely among the states, although there is some agreement on such basic issues as which forms of life are considered game and which are not. So varied are these laws that they resist easy summary. Still, a number of legal issues tend to arise repeatedly. These are the issues addressed in this chapter.

It is important to begin by noting that state fish and game agencies are often directed by an appointed or elected group called a commission or a council. This body makes the policy that the staff of the agency implements. Some of the recurrent issues arise in relation to the commissions and some in relation to the combination of commissioners and agencies. When the legal question concerns only the commissioners, we will speak of "commissions"; when the question concerns the combination of commissioners and agencies, we will speak of "agencies."

We first look at the agencies that states create to promulgate detailed fish and game regulations and to implement them season by season. The creation and empowerment of these agencies give rise to a number of recurring legal and policy questions. Although fish and game agencies are staffed with people trained in wildlife management and science, commission members generally are chosen based on other criteria. Can selection methods rule out people who do not support hunting and fishing? What powers do the commissions and the agencies they oversee possess to designate game species, to set rules on methods of capture, and to prescribe different capture rules for different places? Finally, what processes must agencies follow as they go about issuing regulations, defining crimes, and setting enforcement policies?

From agencies and their processes, we'll turn to a number of recurring questions that relate to legal definitions of wrongful conduct. As we've noted, enforcement officers face unusual obstacles in proving violations of fish and game laws. Violations tend to occur in private. It is hard to know and to prove what a person intended when pulling a trigger or releasing an arrow. Legislatures and administrative bodies have often defined criminal conduct quite broadly, making it easier for enforcement officers to prove their cases. But criminal violations that are defined broadly and vaguely raise a host of constitutional questions. Aside from these constitutional issues, there are several recurring issues having to do with the ways specific offenses are defined and applied. When exactly is a person "hunting," for instance? What activities qualify as hunting over bait, spotlighting, and the waste of meat? And what issues arise regularly when lawmakers try to define criminal fish and game offenses in terms of the knowledge or intent of a person accused of a violation? We end this part of the chapter by addressing two particular issues inherent in the law enforcement process itself: the legality of highway checkpoints set up by game officers and the search-and-seizure law that applies in open rural areas.

State law also largely covers another major component of wildlife law, the issue of harm caused by wildlife and whether parties injured by animals should be able to obtain relief. This issue is the flip side of one already considered, whether users of wildlife can get legal relief when their wildlife activities are harmed by other people. Wildlife can cause harm to both people and their private property. Hardly any group of wildlife cases is more fascinating to read, albeit gruesome and sometimes deadly. When animals and people come into conflict, people often end up on the losing side. A useful way to approach the law here is to divide the cases into categories: cases relating to *property damage* versus *personal injury*, and those relating to *privately owned* wildlife versus *publicly owned* wildlife. All are taken up in the chapter's final section.

FISH AND GAME AGENCIES

States differ widely in the number and types of agencies they create to regulate hunting, fishing, and wildlife. Some states have one agency; others have more. The variations have much to do with the relative importance of different species in different states. Predictably, states with significant commercial fisheries are more likely to have a separate agency for fish than those that have little commercial fishing.

Agency Composition

A common complaint about state fish and game commissions over the decades—almost since they first arose a century ago—is that they are controlled by people strongly supportive of hunting and fishing who are inclined to manage wildlife for the particular benefit of hunters and fishers. The problem, or at least the reality, is that the composition of commissions is sometimes specifically limited by law to people who hunt or fish, or to representatives of particular groups or constituencies. Outsiders have criticized these selection criteria and processes. Few animal welfare advocates, for instance, got appointed, at least until the past decade or two. Both trained wildlife scientists and organized environmental interests have also been left out. When commissions are staffed this way, critics contend, they sometimes keep game populations unduly high, leading to environmental degradation. They excessively favor game populations over nongame species.

State legislatures, it seems, face few legal limits on the ways they structure their commissions, even when a policy slant is manifest. For example, in a prominent 1976 New Jersey ruling, the Humane Society and the Sierra Club challenged the appointment system to the Fish and Game Council, claiming violations of equal protection and due process.[4] The governing statute provided for an eleven-member council, appointed by the governor with the consent of the state senate. Three of the eleven members had to be farmers recommended by an agricultural convention, six were sportsmen recommended by the state federation of sportsmen's clubs, and two were commercial fishermen. The individual plaintiffs in the case, all knowledgeable about wildlife, enjoyed the state's wildlife "for purposes other than hunting, fishing, and farming." The state supreme court subjected the statute to a highly deferential, rational-basis scrutiny. In the court's view, the statute rested on a rational basis. The prominent nominating role given the private groups did not offend due process. This New Jersey decision is similar to ones in other states. Courts have been unwilling to tackle what they plainly view as an issue better suited for political resolution.

Unconstitutional Delegation

Another line of cases challenging state agencies are those claiming that legislatures give them too much discretion in regulating activities as they see fit. The chief complaint is that this work is essentially legislative in nature. It is work that the legislature itself should undertake, rather than delegate to an administrative entity. Constitutionally, legislatures cannot delegate their lawmaking function to someone else. Many legislatures, however, have given agencies scant guidance on how they should go about exercising their considerable delegated powers. At what point does broad discretion become an abdication of legislative function?

For the most part, courts seem comfortable with these legislative arrangements despite the breadth of agency discretion. Courts require only that legislatures supply a statement of policy, or an indication of an overall purpose or goal, for the agency to use to guide its actions. In a 1997 ruling, for instance, the Supreme Court of Tennessee upheld a statute that empowered the state Wildlife Resources Commission to add and delete animals from a legislatively drafted list of dangerous species.[5] The statute said nothing specifically to the agency about how it should exercise this power. Nonetheless, the court that assessed the statute viewed the legislature's guidance to the agency as sufficiently clear. The legislature itself provided a list of dangerous species; that list, the court said, implicitly provided guidance on the types of traits that would make other species appropriate for the list. In addition, the court interpreted the statute to require that the agency act consistent with a "standard of reasonableness." An even broader delegation of power was upheld by the South Dakota Supreme Court in 2004.[6] The state's agency wielded extensive power to define wildlife crimes, including felonies. The court upheld the grant of power even though the statute provided, by way of policy or legislative intent, nothing more than a generalized statement about the need to protect the state's wildlife.

Exceeding Delegations of Power

More common than the claim of unconstitutional delegation of power to an agency are assertions that agencies have acted unlawfully by exceeding the specific powers vested in them by the legislature. A 2004 ruling by the Wisconsin Supreme Court explained the basic law on the issue:

> In determining whether an administrative agency exceeded the scope of its authority in promulgating a rule, we must examine the enabling statute to ascertain whether the statute grants express or implied authorization for the rule. It is axiomatic that because the legislature creates administrative agencies as part of the executive branch, such agencies have "only those powers which are expressly conferred or which are necessarily implied by the statutes under which it operates." Therefore, an agency's enabling statute is to be strictly construed. We resolve any reasonable doubt pertaining to an agency's implied powers against the agency. (citations omitted)[7]

The Wisconsin case involved a type of dispute that has arisen often: whether a particular state agency has the power to add species to the game list and then set hunting seasons on them. The dispute involved a bird that has appeared often in such disputes, the mourning dove. The relevant Wisconsin statute authorized the agency to set open and closed seasons for game, and otherwise to impose hunting limits, to "conserve the fish and game supply and ensure the citizens of [Wisconsin] continued opportunities for good fishing, hunting and trapping." The statute defined game as "all varieties of mammals or birds." As thus written, according to the court, this statute gave the agency the power as it saw fit to add or delete any mammal or bird from the game list. Given this delegation of broad power, the agency was within its authority to set hunting seasons for mourning doves. The outcome of the case depended, of course, on the exact language of the Wisconsin statute. In other states, agencies lack this power because legislatures have retained it.

A related question involves the authority an agency possesses over animals born and raised in captivity. Agency powers tend to vary when it comes to animals that straddle the line between wild and domestic or that are raised in captive, farm-type operations for sale to game ranches. The uncertainty over agency powers can extend also to game ranches themselves, and to similar operations in which landowners breed and raise animals for use solely on their own lands. What regulatory power do agencies have over these animals and operations? The answer, again, varies from state to state, and a careful review of state laws is needed to identify agency powers. A typical dispute was resolved by the Supreme Court of Oregon in 2006 in *State v. Couch*, discussed in chapter 1.[8] A private landowner claimed that the state Fish and Wildlife Commission had no authority over the fallow, axis, and sika deer that he imported from outside the state. The deer were not indigenous to Oregon and never lived in the state outside captivity. According to the landowner, the agency had power only over "wild" animals, and these deer were not wild. To resolve the case, the court carefully reviewed the statute, noting that "'wildlife' means whatever the legislature says that it means." The legislature's definition included, as wildlife, all wild mammals. The court interpreted this as applying to all deer, whether or not native and whether or not confined.

Disputes about agency power have taken many forms, all variations on the basic question of what authority the agency has received from a legislature. Indeed, the factual permutations of the disputes are quite numerous. In a 1998 dispute, for instance, a Washington court had to decide whether or not an agency's power to regulate the "manner of taking" game included the power to require that hunters wear orange (the court said yes).[9] A 1995 case from Massachusetts considered whether a state agency could authorize the used of "padded jaw traps" by trappers, given that the legislature had expressed banned the use of "steel jaw leghold traps" (again, the court said yes).[10]

Local or Special Laws

A variant on these delegation-of-power cases is the various disputes that call into question the power of wildlife agencies to set different rules for different places within the state. Agencies typically possess this power and exercise it freely, but the legality of their work is not always free from doubt. Particular rules can be deemed arbitrary, even when an agency has the power to draw geographic lines. Generally, rules that differ among parts of a state are subject to review under a deferential standard that asks only whether they are supported by a rational basis—or, even more deferentially, whether they are so obviously misguided as to seem arbitrary and capricious.

Many state constitutions ban state lawmakers from enacting "special" or "local" laws that discriminate against parts of the state. The problem of local laws was acute in the late nineteenth century, when many of these constitutional provisions were adopted. Courts have almost always rejected claims that wildlife laws violate these state constitutional provisions. An Arizona court, for instance, heard such a dispute in 1999.[11] A state statute banned the taking of wildlife on public lands using leghold traps but allowed the traps elsewhere. The court rejected the idea that the challenged statute amounted to special legislation because it applied to all citizens equally and served a legitimate governmental purpose related to safety and conservation. The statute was also upheld under the equal protection clause because it was supported by an adequate rational policy basis.

Following Procedures

A final group of challenges to agency actions is those contending that an agency has failed to follow the proper procedures when it issued a regulation or took some other action. Generally, agencies are held to strict account when it comes to procedures. If a mandatory procedure is skipped, the agency action is set aside and the agency told to redo the work. An exception usually exists for procedural violations that are harmless in that the violation did not hurt anyone.

A number of states require agencies to perform environmental studies of one type or another before taking action that can affect the environment. These rules and other express study requirements sometimes apply to wildlife agencies. When they do, the agencies must follow them or courts will strike down their actions. The most difficult cases of this type tend to be those in which the governing statute contains vague statements about scientific study or other forms of assessment—not mandating a particular process, but nonetheless demonstrating an expectation that agency action would build upon a study effort of some sort. Is it possible to challenge an agency action when the statutory process is so vague?

An illustrative case arose in Oregon in 1988 concerning a proposed hunting season for cougar.[12] The statute gave the Oregon Department of Fish and Wildlife the power to prescribe open seasons "after investigation of the supply and condition of wildlife." Did this language require a scientific investigation or other formal study, or was it enough for the agency to rely on already available data? The challengers to the agency's action contended that the agency acted with little real information about cougar populations and conditions. In the end, the court upheld the agency's open season, even though the state had apparently never conducted a comprehensive study of the animal. The agency acted instead based on a count of the number of cougar taken and a record of complaints of cougar damage, along with general knowledge of cougar behavior. In the court's view, this inquiry was sufficient to satisfy the vaguely worded state statute.

DEFINING WRONGFUL CONDUCT

One of the chief tasks of legislatures and wildlife agencies is to prescribe rules of conduct and to set penalties for violations. As hunters and fishers well know, statutes and regulations often prohibit a wide variety of conduct, sometimes in minute detail. As lawmakers go about their work, they encounter basic challenges. Two of them are explored here: fairness (vagueness, overbreadth, and definitional issues), and constitutional limitations on prosecutions for violating fish and game statutes.

Fairness Versus Enforceability

One challenge arises out of the difficulty of enforcing wildlife laws when the conduct at issue is private and much of the information about it cannot be known or proved with certainty. In criminal prosecutions, the state must prove all elements of an offense beyond a reasonable doubt. This is a substantial burden of proof, the highest in our legal system. How can prosecutors meet this burden when they have no witnesses to the conduct? And how can they prove what a person knew, or what a person intended, without gaining access to a person's hidden thoughts? Was a person walking down a path with a rifle engaged in hunting or merely going for a walk? Did a hunter know that she was hunting near bait? Was a person driving slowly along a road at night stalking deer? Did the person who shot a wolf realize that it wasn't a coyote?

If lawmakers were concerned only about drafting fair laws—laws that distinguished carefully between conduct that seems bad and conduct that does not—they would likely define crimes to take into account a person's knowledge and intent. They would distinguish clearly and crisply between wrongful action and action that is not wrong. The problem with this approach is that precise, detailed rules create too many enforcement problems. Prosecutors often cannot prove intent or knowledge, and behavior is sometimes ambiguous. Lawmakers have responded to enforcement challenges by defining criminal conduct more broadly than they otherwise might, so as to make prosecutions much easier. In the case of many offenses, they define crimes in terms of circumstantial acts and often omit considerations of knowledge and intent. In doing so, they define wrongful conduct in ways that also penalize behavior that could be entirely innocent. On the positive side, these simple, broad definitions of crimes make prosecutions easier. Wrongful conduct is therefore more likely to incur punishment. On the negative side, though, broad definitions can lead to convictions of people who really were not acting wrongfully. Prosecutorial discretion thus becomes especially important. We must trust prosecutors not to bring cases when innocent behavior happens to run afoul of a broadly written criminal provision.

The bottom line is this: a trade-off exists between fairness and enforceability, a trade-off that is always present even as lawmakers struggle to find the appropriate balance. To a large extent, the fairness-enforceability trade-off is left for resolution by legislatures and wildlife agencies, which do the primary work of drafting criminal provisions. Courts, though, also get involved when they hear prosecutions. Ultimately, courts decide the guilt or innocence of a defendant. They pass final judgment on whether laws are written in ways that comply with the Constitution and with basic notions of fairness. Fair laws, for instance, must provide clear notice to people about what is and is not prohibited. Fair laws punish only the behavior that the legislature sought to condemn, not innocent behavior.

Avoiding Unfairness

As courts consider wildlife prosecutions, they have several tools for dealing with these challenges, thereby interjecting fairness when laws seem to lack it. One tool they use is their power to interpret particular criminal statutes. They sometimes do so in ways that narrow the application of statutes so that they do not apply to seemingly innocent behavior. They can, for instance, interpret particular words narrowly so as to avoid problems (interpreting "animal" to mean only "game animal" to avoid application to nongame species). In addition, they can interpret a criminal statute as requiring some showing by the prosecution that the defendant had knowledge of particular facts (for instance, that bait was in the area) or was motivated by some particular intent (for instance, intent to take game), even when the statute does not seem to require that the prosecutor prove these elements. The effect, again, is to tailor the circumstances in which a statute applies to avoid penalizing behavior that doesn't seem to warrant punishment. When courts take these steps, however, they arguably breech their duties to enforce the laws as written. Legislatures are the ones who define criminal conduct. If the legislature wants to cast its net broadly, penalizing conduct that doesn't seem to warrant it, perhaps the solution is to go back to the legislature and propose changes rather than let courts in effect revise the statutes as they deem wise.

Vagueness and Overbreadth

As a matter of constitutional due process, a criminal statute must provide a fair degree of notice to citizens about the conduct that is prohibited. Also, it cannot sweep so widely, in terms of the behavior it prohibits, as to include conduct that is clearly not wrongful. On both questions, the issue is one of degree. How clear must a statute be? And how precisely must it be written so as to penalize only wrongful behavior? The issues are intertwined, and both are related to the challenges states face in drafting laws that can be enforced, given the difficulties of proof that prosecutors face. A 2004 Idaho ruling explained the basic rationale:

> The void-for-vagueness doctrine is premised upon the Due Process Clause of the Fourteenth Amendment. This doctrine requires that a statute defining criminal conduct be worded with sufficient clarity and definiteness to permit ordinary people to understand what conduct is prohibited and to prevent arbitrary and discriminatory enforcement. An enactment is void for vagueness if its prohibitions are not clearly defined. Due process requires that all be informed as to what the state commands or forbids and that persons of common intelligence not be forced to guess at the meaning of the criminal law. (citations omitted)[13]

Recent judicial rulings on these constitutional doctrines illustrate the practical difficulty of applying them. Not surprisingly, courts have reached different conclusions on essentially identical facts. As shown in the following examples, some courts protect defendants by insisting on narrowly drawn statutes. Others side with law enforcement officials and their perceived need to define statutes broadly.

In 1997, the Arkansas Supreme Court considered a typical state statute banning road hunting.[14] The questions: when are people driving down a road hunting, and when instead are they simply driving and watching for wildlife? Arkansas lawmakers chose to define road hunting broadly to facilitate prosecution. Its criminal code provided that a prima facie case of road hunting was shown whenever a person, during deer hunting season, drove down a road in which wild game was likely to be present with a loaded or uncased firearm anywhere in the vehicle. The court ruled that the statute was unconstitutionally overbroad and thus invalid. The wording of the statute was "so inclusive that it may affect the rights of non-hunters who possess loaded or uncased firearms" to use public roads. The statute improperly "include[d] within its sweep innocent and legitimate conduct."

In a 1994 ruling, the Washington Supreme Court considered challenges to the state statute prohibiting "spotlighting"—that is, using artificial light after sunset to hunt deer and other big game.[15] The statute banned the use of artificial lights to hunt big game. It then provided that a prima facie violation was shown whenever a person was found with a spotlight or other artificial light, and with a weapon, in a place where big game might reasonably be expected. The headlights of a vehicle qualified as artificial light. The defendant challenged the statute as both vague and overbroad, claiming that it penalized simple driving at night with a weapon in a vehicle. The court upheld the statute by interpreting it to require a showing that the defendant was actually hunting, which the defendant, on the facts, was clearly doing.

In a 1997 ruling, the Kansas Supreme Court struck down as overbroad a state statute that made it a crime to cast light "for the purpose of spotting, locating or taking any animal" while possessing a gun.[16] The statute was defective, the court ruled, because "animal" could include livestock owned by the defendant. The defendant was allowed to raise this legal defense without any showing that he was in fact looking for livestock.

In a nearly identical 1983 case, the South Dakota Supreme Court ruled to the contrary that a defendant could not raise this legal defense unless he was in fact looking for livestock.[17] The court nonetheless interjected fairness into the statute by interpreting the term "animal" in the statute to mean "game animal," thereby excluding people who were in fact looking for livestock.

In a 1986 ruling, a Kentucky court struck down as overbroad a spotlighting statute that required no showing of an intent to hunt but penalized merely casting light while possessing a firearm.[18] The defendant was allowed to challenge the statute even if he was actually taking game by artificial light.

In a 2004 Idaho decision, the court ruled that a statute was not vague when it imposed added punishments for unlawfully killing and wasting a "trophy deer."[19] The statute defined this term by reference to standards established by the national Boone and Crockett Club, which had no local chapters. The club standards in turn required application of a complex measuring system that entailed a certain amount of subjective judgment.

Nonetheless, the term was adequately clear, the court announced, because the measuring method was explained in a manual that could be obtained through bookstores or through the Boone and Crockett Club.

In other rulings, courts have concluded: that the term "bait" in a hunting-over-bait statute is not unduly vague;[20] that a statute prohibiting takes of "illegal deer" is not overbroad because it applies to an accidental killing of a doe;[21] and that the term "driving deer" is not overly vague.[22]

What Is Hunting?

Perhaps no issue has engendered more litigation than the definition of hunting and the challenge of defining the behavior that fits within it. Hunting, of course, doesn't mean capturing or killing animals, or even getting a shot at them. It includes merely searching. But when is a person searching with an intent to take animals, and when is a person simply enjoying nature and wildlife viewing? The issue arises under countless statutes and has given courts no small number of headaches. Again, the following survey of recent rulings helps illustrate the challenge for lawmakers, prosecutors, and hunters themselves.

In a 1997 ruling, an Idaho court agreed that a man was hunting when he walked down a forest path dressed in camouflage clothing, carried a compound bow and arrows, and had an elk bugle call in his mouth.[23] The prosecutor did not have to show that he was engaged in chasing, driving, attacking or pursuing elk.

A Louisiana court in 1995 had occasion to consider the definition of hunting in a statute that banned hunting deer during illegal hours.[24] The defendants, intoxicated, were driving at night with two high-powered rifles in the front seat, obviously looking for deer. Their vehicle also contained hunting clothes and camping equipment, but no spotlight. The defendants claimed they were merely looking for deer to hunt the next day. The prosecution failed to offer evidence as to whether the rifles were loaded or whether the vehicle contained ammunition. The court reversed the defendants' convictions because the prosecution had not excluded reasonable doubt as to whether the defendants were, as they claimed, simply looking for deer to hunt later.

The Washington Supreme Court in 1994 considered whether a defendant could claim he was not hunting big game when he got out of his truck and fired his rifle at a decoy deer set out by game enforcement officers.[25] The trial court ruled that the defendant was not hunting because it was factually impossible to kill game by shooting at a decoy. The high court reversed the ruling. The prosecution did not have to prove that the defendant actually encountered big game, the high court stated. It was sufficient that the defendant "made an effort to kill or injure big game in an area where such animals may reasonably be expected." The prosecution proved its case by showing that the defendant took aim; it was not necessary to show that he actually fired a shot.

In a 2002 ruling, a Kansas court considered whether two defendants were engaged in unlawful hunting with a spotlight when they shone a spotlight on a deer from their vehicle, even though they had no weapons with them.[26] A second vehicle, driving with them, saw the spotlighted deer and then drove off the road in pursuit of it. The statute defined as hunting any use of a spotlight by a person who had in his possession "any rifle, pistol, shotgun, bow or *other implement* whereby wildlife could be taken." The statute defined "take" to include harass, pursue, and molest; it was not limited to shooting, wounding, and killing. Held: the defendants in the vehicle that did the spotlighting were guilty of unlawful hunting. Although they did not have possession of a rifle, pistol, shotgun, or bow, the vehicle they were driving itself amounted to an "implement whereby wildlife could be taken," given that a taking occurred simply by chasing an animal.

Finally, in a 2004 ruling, a New Jersey court decided that a hunter who had killed and tagged a deer early in the first day of hunting season was not hunting later when he removed his shells from his shotgun and gave them to another hunter, put his gun in a cloth case, but then carried the gun with him as he watched other hunters rather than leaving the gun unattended in his truck.[27] The shotgun was not "readily usable," the court ruled, when the hunter had no shells, the gun was cased, and no other hunters were within fifty yards.

Hunting Over Bait

Just as contentious as the definition of hunting has been the crime of hunting over bait—the practice of putting out bait to lure game animals for easy shooting—and the recurring question of whether the prosecution must prove that the hunter-defendant had actual knowledge of the bait. This offense offers a stark illustration of the inevitable trade-off between fairness and enforceability. On the one side, it seems unfair to convict a hunter of the offense when the hunter really did not know of the bait and perhaps had no reason to know. On the other side, it is difficult for prosecutors to prove knowledge beyond a reasonable doubt when a hunter claims he did not know. A hunter who claims ignorance could almost always avoid liability.

The typical approach for some decades was for lawmakers to define the offense so as to require no showing of knowledge, and then leave it to prosecutors to refrain from prosecuting cases where the hunter really did not know about the bait. That approach continues to be followed in some states. A rising trend, though, has been to require some showing by the prosecution either of knowledge of the bait or of a negligent failure by the hunter to be aware of the nearby bait.

A ruling showing the range of approaches on this issue was handed down by a Pennsylvania court in 2000.[28] The applicable state statute required no particular knowledge or mental state on behalf of the defendant; it defined the crime simply as hunting over bait. The prosecution took the view that this was a "strict liability" crime in the sense that no knowledge or intent was required: the mere act of hunting with bait nearby violated the statute. The defendant, on the other side, claimed that the state should have to show a specific intent on behalf of the hunters to make use of the bait in hunting. The court rejected both approaches. The court, in effect, added to the statute a requirement that it termed the "reasonable hunter" or "negligent hunter" test. A person violated the statute if and only if she hunted with knowledge of the bait or hunted under circumstances in which a reasonable hunter would have been aware of the bait. The court was unwilling to interpret the statute as a strict liability offense, given that "a completely innocent hunter, exercising the utmost of good citizenship and sportsmanship might unknowingly hunt in a baited area." It was too great a burden on the prosecution, however, to prove not just knowledge of bait but an intent to take advantage of it.

Mental State in General

In recent years, more and more lawmakers have shown unease at the notion of game law violations that require no showing of knowledge or intent. Various rulings from Oregon, Alaska, Wisconsin, and elsewhere have interjected requirements that prosecutors make some showing of culpable mental state—usually some form of "knew or should have known" standard—in order to obtain convictions. On the other hand, many offenses are plainly written as strict liability rules, and courts continue to enforce them as written. Thus, for instance, courts commonly reject claims by hunters that they made mistakes about the animals they shot. The Vermont Supreme Court in 1999 (to use one of many examples) refused to allow a hunter to escape conviction for shooting a moose out of season by claiming he thought the animal was a deer.[29] The offense required no such knowledge of the animal, the court concluded. Similarly, an Ohio appellate court in 2002 confirmed that no mental state was required for the crime of hunting deer without possessing a valid license.[30] The defendant inadvertently left his license in a vehicle two miles away. The court ruled that the prosecution had no duty to show recklessness or even simple negligence in failing to carry the license.

Waste

A final crime that might be mentioned, by way of illustrating the types of legal issues that arise, is the offense of wasting game, an offense committed when a hunter kills an animal and fails to carry away the edible portions of the meat. The rationales for the statute are several and have apparently shifted over time. An early rationale reflected popular disapproval of hunting practices in which hunters slaughtered game simply for the pleasure of it, leaving the meat to rot. Criminal penalties diminished that problem. Waste statutes may have served to conserve large-game-animal populations by reducing the number of animals a hunter could take in one day. They may have reflected, too, a disdain for big-city hunters who invaded rural hunting grounds and who were not in a position to pack meat to take it home.

A more modern rationale is that waste statutes aid enforcement of game laws generally. Many violators of game laws are interested in obtaining valuable parts of particular animals (for instance, black bear gall bladders or big-game horns). Violators may have little interest in the remaining parts of the animals, which in any event could be difficult to remove and which, when transported, could provide evidence of game law violations. Abandoned animal carcasses provide prima facie evidence of a game law violation, without any other required showing. In a 1992 ruling, the Montana Supreme Court concluded that its state ban on wasting game was a strict liability offense, requiring no showing of culpable mental state.[31] The defendant in the case claimed that he was unable to retrieve the meat from a mountain goat, killed in a remote location, until a day after the hunt, at which point the meat was spoiled. Other hunters in the area disagreed with the factual assertion. The defendant was not allowed to claim that he had no intent to commit waste, nor did he succeed in claiming that the statute was unconstitutionally vague. In a 2007 ruling, a Colorado court held that a defendant violated a statute penalizing the killing and abandonment of wildlife so long as he knew that he was abandoning it.[32] The prosecution did not need to show a specific intent by the hunter to abandon.

LIMITS ON LAW ENFORCEMENT

Game law enforcement is hardly less simple than the enforcement of other laws. The applicable constitutional provisions, mostly designed to protect criminal defendants, are numerous and complex, deserving of extended study on their own. Two of the many relevant issues stand out as particularly pertinent to game law enforcement: traffic checkpoints and open-field searches.

Occasionally game enforcement officers find it useful to set up traffic stops to check for unlawful game and for hunting licenses. Such stops raise difficult constitutional issues because they amount to searches and seizures that typically are possible only in accordance with a search warrant issued by a court based on a showing of probable cause to believe that the search will produce evidence of a crime. Emergency exceptions exist, of course, in which warrants are unneeded, generally because time does not permit. But traffic stops are planned in advance, and timing is not an issue. What justifies warrantless searches under such circumstances?

Law on the subject is somewhat unclear. The U.S. Supreme Court has not had occasion to consider game law checkpoints in particular. Since the Fourth Amendment only prohibits "unreasonable searches and seizures," the threshold question is whether checkpoints are unreasonable. As a result, lower-court rulings typically evaluate the legality of such checkpoints based on an assessment of several factors that focus on reasonableness: the magnitude of the state interest served by the checkpoint; the extent to which the checkpoint is effective in serving that state interest; and the degree of intrusion involved in the search. Some courts have viewed game laws as relatively unimportant and have assigned the state interest a low value; other courts have concluded that the state's interest in wildlife conservation is, in fact, quite important.

It is clear that checkpoints must be conducted in accordance with a specific agency policy and approved by supervisors at a relatively high level. It is not lawful for individual officers to decide to set up checkpoints. It is helpful also, in terms of sustaining the checkpoint's legality, to set up the checkpoint at a time and place in which most vehicles being stopped will include hunters or fishers. Simply looking inside vehicles, with or without a flashlight, is less intrusive than conducting a physical search of the vehicle. In addition, game law checks are more likely valid when the officials operating them are only looking for game law violations and do not attempt to use the stop to conduct searches for a wide array of criminal activities. Vehicles should not be stopped based on "hunches," but rather on established criteria (for example, every vehicle, every fourth vehicle, and so forth); the less discretion agents have, the better. One danger of searches is that they take people by surprise and can scare them. Game checkpoints should therefore be very clearly marked with signs and/or flares, with official vehicles present, so that drivers of vehicles being stopped know immediately who is stopping them and perhaps why.

A 2006 federal court ruling from North Carolina provides evidence of the limits of this exception to the otherwise blanket need to obtain search warrants.[33] A U.S. Forest Service ranger and a state wildlife officer set up a checkpoint on a gravel road that allowed travel in only one direction due to a recent landslide. The federal ranger was operating on the assumption that the checkpoint was conducted in accordance with state wildlife policy; it had not been approved in accordance with U.S. Forest Service regulations. The state game officer stated that he set up the checkpoint at the instruction of superiors, but he could not state whether it was conducted under the written policy of the state wildlife commission and did not in court produce a copy of the state's policy. He did not know whether he was required to post signs informing the public of the purpose of the checkpoint, his vehicle was unmarked, and no supervisor was present. Finally, the checkpoint was set up on a Sunday, a day in which hunting was prohibited under state law. Based on all the factors, the court concluded that the checkpoint violated the Fourth Amendment ban on searches and seizures.

Aside from checkpoints, game officers with adequate enforcement authority can stop a vehicle driver for questioning if they have a reasonable, articulable suspicion that the person is engaged in criminal activity. In addition, the ban on warrantless searches applies only when a defendant has a reasonable expectation of privacy. This limitation is often important when game officers check fishing boats for violations. Officers are free to look in open boats whether on the water or on shore, given that the boats are exposed to public view. They can also check inside boats in places where fish are typically kept to see what fish, if any, have been caught. The rationale for this result was explained by the Montana Supreme Court in a 2002 ruling:

> In engaging in this highly regulated activity, anglers must assume the burdens of the sport as well as its benefits. Thus, no objectively reasonable expectation of privacy exists when a wildlife enforcement officer checks for hunting and fishing licenses in open season near game habitat, inquires about game taken, and requests to inspect game in the field. In this capacity, game wardens are acting not only as law enforcement officers but as public trustees protecting and conserving Montana's wildlife and habitat for all of its citizens.[34]

In the case of citizens not in vehicles or boats, game officers can stop them, ask to see hunting licenses, and inquire about game, at least when the people are stopped in or departing from hunting areas and when the particular people being stopped are not singled out. The Louisiana Supreme Court explained the need for this power in a prominent 1994 ruling: "If agents were not entitled to make suspicionless stops of hunters for license checks and brief questioning, the enforcement of the game limit laws would be retrenched to an unacceptable level."[35]

Far more settled is the constitutional law dealing with game officer searches of rural lands for evidence of game law violations. Federal constitutional law is rather clear in authorizing game officials to enter and look around land that is not close to a private dwelling—land that is not, as the Supreme Court has put it, within the immediate "curtilage" of a home. The extent of a home's curtilage is sometimes hard to discern, and is based on an assessment of several factors. Outside that area, however, the landowner has no expectation of privacy and thus no power to complain about searches of the land. This is true even if the land is posted against trespassing. This rule of law is usually termed the "open field" doctrine, though the name is a bit inapt. The land need not be physically open—it can be fenced—and it can be a forest, a wetland, or even a lake rather than a field.

Landowners who have resisted searches of their land have had modest success in a few states in getting state courts to rule that such searches violate the search-and-seizure bans contained in their *state* constitutions, even though they do not violate the *federal* Constitution. In a 1999 ruling, the Montana Supreme Court excluded evidence gathered by state agents who entered the defendant's rural land, which was posted against trespassing and largely screened by brush from public sight.[36] The Mississippi court a year earlier had similarly ruled that game wardens could not enter land posted against trespassing.[37] It also concluded, though, that only the landowner could raise the objection—not others—and wardens could enter if they observed evidence of a violation before entering. The more typical result was reached in 2007 by the Pennsylvania high court, which ruled that its constitution offered no greater protection than the federal Constitution.[38]

A more difficult case is presented when game officers do more than simply walk and look around private land. In a 1999 ruling, the Arkansas Supreme Court decided that game officers had authority to climb up a ladder and look in a deer stand that was essentially a metal box "with sides three to four feet high and a roof elevated on poles at each corner of the box."[39] Due to the openness of the stand, its owners had no expectation of privacy. In contrast, the Minnesota Supreme Court in 2002 decided that a fisher had an expectation of privacy covering a fully enclosed icehouse used for ice fishing.[40] A warrantless search of the icehouse was unlawful.

WHEN WILDLIFE CAUSES HARM

For generations, courts have heard cases in which plaintiffs have been harmed by wildlife and have sought to recover damages from animal owners. The rulings deal with three types of disputes, relating to personal injuries, property damage, and harms caused by government.

Personal Injury

Many cases dealing with wildlife harm have involved personal injuries suffered by the plaintiffs. In these cases, a rather sharp line is typically drawn between animals that are naturally ferocious or dangerous and animals that are not. When an animal that is considered dangerous causes personal injury, the owner is likely to be found liable. This distinction based on dangerousness raises obvious problems, given the variety of exotic animals that people keep as pets. Is a monkey, kept as a house pet for more than a decade, still a dangerous animal, or has it become safe? What about trained animals that work in zoos? Other issues also arise. What happens when a wild animal escapes and the owner's property rights in it come to an end? Is the owner nonetheless still liable for the harm? Finally, what about wild animals that live on privately owned land, particularly when the owner knows about the animals and recognizes their dangers? Is a landowner liable if a visitor to his land is injured by a wild animal? Does it make a difference whether the landowner is running a business and the visitor is a customer? Does the landowner have a duty to warn of natural dangers?

The law applicable to personal injuries—"tort" law—recognizes three types of conduct that result in the actor's liability to the person who is injured: intentional conduct, negligent conduct, and conduct that is subject to strict liability—that is, liability without intentional or negligent acts. The longstanding common law rule is that the owner of a dangerous animal is strictly liable for the personal injuries that the animal causes. In the dangerous animal case, due care is not a defense. Even careful animal owners can be held liable. The rationale for this strict approach to liability was explained in a federal appellate court ruling in 1995:

> Keeping a tiger in one's backyard would be an example of an abnormally hazardous activity. The hazard is such, relative to the value of the activity, that we desire not just that the owner take all due care that the tiger not escape, but that he consider seriously the possibility of getting rid of the tiger altogether; and we give him an incentive to consider this course of action by declining to make the exercise of due care a defense in a suit based on an injury caused by the tiger—in other words, by making him strictly liable for any such injury.[41]

Thus the reason an animal owner faces liability is not because the owner failed to act carefully, but because the owner introduced the animal into a place where it posed a danger to other people. Having created the danger, the owner is liable for any resulting harm. This liability, however, is limited to harm that is caused by the animal's dangerous tendencies. If the harm does not arise due to the dangerous tendency, then the more common standard of negligence will apply.

This legal approach to liability is typically referred to as "strict" liability, to distinguish it from liability based on fault—that is, based on either negligence or intentional misconduct. Strict liability, however, is not absolute liability. There are reasons why a person injured by a dangerous animal would nonetheless not be able to recover compensation for his injuries. If the injured person himself is at fault, causing or contributing to the harm, recovery can be barred. For instance, trespassers on land are often unable to recover compensation for harm done by dangerous animals. A more general exception to strict liability is for a person who "knowingly and unreasonably" subjects himself to the risk of harm from a wild animal. A common instance here would be a person who approaches a wild animal that is chained and, for no good reason, knowingly gets close enough for the animal to harm him. This exception to recovery, however, is a fairly narrow one. A person can recover compensation for an animal-inflicted injury even if he is careless—that is, even if he fails to exercise ordinary care in observing the presence of an animal and in escaping from its attack. Recovery is barred only if a person knows the harm and deliberately and without good cause subjects himself to it.

A 1997 Indiana ruling illustrates these principles.[42] The case involved an animal breeding center that raised a variety of exotic animals, including Siberian tigers. A man in his twenties regularly visited the place and on occasion saw people petting the tigers through a fence. The man, a friend of a tenant on the property, was allowed to roam the property freely. On the day he was injured, the man had consumed considerable alcohol. He reached into the fenced area to pet a male tiger. While doing so, a female tiger "made some commotion" and distracted the man's attention. While he was looking away, the male tiger pulled the man's arm into the cage and mangled it. According to the court, the man could recover compensation unless he "knowingly and unreasonably put himself within reach of a wild animal that was effectively chained or otherwise confined." Phrased differently, the issue was whether he acted with "a mental state of venturousness and a conscious, deliberate and intentional embarkation upon the course of conduct with knowledge of the circumstances." On the facts of the case, it appeared that the man might have knowingly assumed the risk of injury in this manner, though it was up to the jury to make that factual determination. The case was returned to the trial court to make this determination.

This strict liability rule governing dangerous animals does not usually apply when the person injured has been hired to work with the animals. In that instance, the employer of the person injured is normally liable only under common law negligence rules or under the terms of any worker's compensation statute that applies. Thus in a 1965 case from Oklahoma involving a city zoo, an employee hired to clean out the pit used to house lions was injured when the door keeping the lions out was not securely fastened and two lions pushed their way in, mauling the employee severely.[43] The trial court awarded damages on the assumption that the strict liability rule governing animals applied to the zoo and its employee. The Oklahoma Supreme Court, however, reversed:

> Here, plaintiff was not a spectator at the zoo, but the undisputed evidence is that he was employed specifically to assist in caring for wild animals on exhibit there, that he had undergone an apprenticeship or "training" period, and that he had had some previous experience in working in the lion pit. There can be no doubt in our opinion that the doctrine of absolute liability has no application.

On-the-job injuries are typically covered by worker's compensation schemes, which allow recovery for nearly all workplace injuries without regard for employer fault. Recovery, however, is limited to specific dollar amounts, often much-lower amounts than an injured party could recover in a tort suit. Employers typically strive to get all injuries included within the worker's compensation law; employees try to get them excluded.

Strict liability applies only for animals that are considered dangerous. The issue of dangerousness is a factual one. One of the law's uncertainties is whether animals are considered dangerous on a species-by-species basis or whether the owner of a particular animal can argue that the animal is not dangerous, even though other members of the same species ordinarily would be. One case that raised the issue was resolved by a Texas appellate court in 1977.[44] Injury was caused by a monkey that had been a family pet for twenty-six years. The court found that the animal was sufficiently domesticated that the strict liability rule would not apply. To recover compensation, the plaintiff therefore had to show "that defendants knew that the animal was accustomed to do mischief, or that the defendants committed acts of negligence."

In two situations involving personal injury, a person can be liable for the injury even though he did not own the animal at the time of the injury. One situation involves the escaped wild animal, in which the former owner's property rights have ended. The general rule here is that the former owner remains liable for harm occurring after the escape, at least until the animal has fully regained its liberty and is living in the wild. Cases of this type have been numerous and high in drama. A 1966 Massachusetts case applied the rule to a zebra running at large on the streets of West Springfield, Massachusetts.[45] A Louisiana case from 1959 used the rule when a sixteen-point antlered deer escaped and gored a police officer in the residential section of a city.[46] A similar Georgia case involved a baboon that escaped from a zoo and entered a private car;[47] a Mississippi case involved a monkey that entered a private home and attacked a girl.[48]

The second type of dispute involving unowned wild animals is the case in which a landowner is sued when a visitor to the land is injured by an unconfined wild animal. Sometimes the owner knows that the animal is present. Other times the owner does not. A growing number of states now have statutes that specifically protect landowners against liability when they open their lands without charge to recreational visitors. These statutes typically allow suits against landowners only if the landowners have engaged in intentional or malicious conduct. Even when landowners are not protected by such statutes, however, injured people rarely are able to recover compensation. Indeed, the near-absolute rule among the states is that landowners have no liability for harm caused by wild animals on their lands, even if they know about the animals. In a 1999 Texas ruling, a visitor to an overnight recreational vehicle park was stung more than a thousand times by fire ants and died; the owner of the park was not liable.[49] In a 1995 Alabama case, a plaintiff could not recover compensation after being bitten by a water moccasin in a private swimming pool.[50] In a Texas case from 1977, a grocery store owner was not liable to a customer who was bitten by a rattlesnake while shopping.[51]

Landowners in such cases are typically liable only if they have acted negligently in some way, either creating the danger or failing to warn of it when they both know about the danger and realize that visitors are unaware of it. In practice, injured parties are rarely able to show that negligence has occurred. Thus in a 1989 California case, a visitor to the defendant's vacation home who was bitten and severely injured by a brown recluse spider was denied recovery because the homeowner had no particular knowledge that the spider was in the home.[52] The reluctance of courts to impose liability in such cases is easy to explain. One reason, surely, is that courts do not want to encourage landowners to engage in the wholesale extermination of native wildlife in an effort to make the outdoors safe.

Property Damage

For obvious reasons, courts rarely are presented with cases involving property damage caused by wild animals that are confined. The law on the subject is therefore not particularly clear. Disputes almost entirely involve wild animals that have escaped. These disputes become especially complicated when the animals have lived for a time in the wild and may even have reproduced in the wild. When an escaped animal mingles with native populations and loses its separate identity, courts seem inclined to end the former owner's liability. The hardest cases involve exotic animals, where the wild population is easily traced to a single owner from whom they have escaped. Should a person who imports an exotic animal into a region be liable for the damage it does after escape, without regard for time? The law at the moment has no clear answer.

An interesting ruling exploring these various issues was handed down by the New Hampshire Supreme Court in 1956.[53] It involved an imported population of Prussian wild boars that were introduced into a fenced twenty-five-thousand-acre wild game preserve. Some escaped over the years, reproduced in the wild, and gave rise to a resident wild population. The boars did considerable damage to property in the area. Because the boars had trouble living during winter without being fed, many of them returned through holes in the game ranch fence to eat feed set out by the ranch owner.

The New Hampshire court considered various theories of liability that might apply in the boar case. One of them was the strict liability rule that applied to domesticated animals that strayed. Owners of domesticated animals remain liable for the damage they do until they are captured. The same rule, the court suggested, ought to apply to the Prussian boars. Even though the wild boars and their offspring had been in the wild for fifteen years, they were still easily traced to the game preserve. As for the strict liability rule that applied to wild animal injuries generally, the court was unwilling to apply that to property damage cases. It was up to the legislature, the court said, to decide whether the normal negligence rule of tort law should be replaced by a strict liability standard.

An additional wrinkle in the New Hampshire case arose because of a state statute enacted in 1949, some eleven years after the first boar escapes. The statute, which applied in two counties of the state, provided that anyone suffering injuries from the boars after April 1, 1950, could recover compensation from the ranch owner. The statute essentially imposed a strict liability rule, and applied the rule to damage caused by animals already on the loose. The court nonetheless upheld the statute as a proper exercise of the police power; it was not an illegitimate, retrospective law that imposed liability for past conduct. The boars were still at large, and the ranch owner under the statute was given time to retrieve them before liability began.

The law on escaping animals, both native and exotic, is poorly developed. Yet it is a vital legal issue for landowners engaged in raising exotic animals or operating game ranches. It is also a vital issue for other landowners who import exotic animals for other purposes—exotic fish, for instance, to raise in ponds. What if an exotic fish species escapes into the wild, causing damage to native fish populations? Could the state hold the importer liable for the harm done to the publicly owned wild animals? Might the state be obligated to do so, given its duties as trustee to protect all wildlife? Similar legal issues arise with respect to other exotic species that are introduced into an area and then escape to become pests. At present, the law has few answers.

A final factual pattern involving wild animal damage to property has to do with the landowner who deliberately attracts wild animals to his property by setting out food or otherwise making the property appealing to the animals. At some point, can the landowner succeed too much, attracting so many animals that neighboring landowners have a right to complain about the property damage they cause?

Cases involving abundant animal populations have typically been handled under nuisance law. A neighbor can recover damages for wildlife harm by showing that the animals have caused substantial harm to his use and enjoyment of property under circumstances that make the imposition of the harm unreasonable. Nuisance cases, that is, require proof on two issues: was the harm substantial enough, and was its imposition unreasonable? Typical of such disputes is a North Carolina case from 1955.[54] There a landowner constructed a three- or four-acre pond on his land and took steps to attract geese by scattering feed and placing "lame wild geese" on the land. Within two years, the pond had become the winter home for some three thousand geese. The geese fed on the plaintiff's adjacent land, destroying his crops and fields. Although recognizing that the landowner did not own the geese because they were unconfined, the court ruled that the landowner could still be found liable for nuisance because he violated the general do-no-harm rule that applied to all landowner activities.

The Special Case of Government Liability

As explained in chapter 4, governments are not liable for property damage caused by wild animals simply because the animals are owned by the state in its sovereign capacity. That rule of nonliability applies not just when animals cause damage to land, but also when they cause damage to personal property (that is, property that is not affixed to land). Governments are liable only for activities that amount to torts—that is, for negligence or nuisance—and then only when the government has agreed to be liable under a tort claims statute or some other law that waives sovereign immunity. Generally, governments are not liable for policy decisions and other discretionary actions that they take. They are liable only for ministerial and operational acts, either carrying out a mandatory statutory obligation or implementing policy decisions once they have been made.

Governments have been held liable for personal injuries under two types of circumstances. One arises when government employees act in a way that significantly increases the chance of wildlife injury and then fail either to warn people of the danger or to protect them against harm. Thus the Alaska Supreme Court in 1979 ruled that the state could be liable for its failure to collect garbage at a state-owned roadside area.[55] The state had put out large garbage collection drums, which were used by people in the area. Inexplicably the state stopped collecting garbage from them. The garbage attracted bears, one of which injured the plaintiff. According to the court, the government could be liable for failing to maintain property that it owned by failing to collect the garbage. It could also be liable in negligence generally, on the ground that it "created a dangerous situation, . . . knew the situation was dangerous, and . . . failed either to correct the situation or to warn people of the danger." Even if the state did not create the dangerous situation, the court opined, it still might be liable if it knew of the specific danger and failed to post warning signs.

Generally speaking, governments are not liable for injuries caused by wild animals, even when the animals are known to inhabit an area and to pose dangers. Thus in a 1995 California case, a child mauled by a mountain lion in a state park could not recover for damages from the state.[56] The lion was a "natural condition" in the park, for which the state had no responsibility. This was true as a matter of law even though lion populations had increased due to a state-mandated moratorium on lion hunting. Similarly, a man severely injured by a shark on a public beach in Florida in 1976 was unable to recover against the city of St. Petersburg, even though various public officials knew that sharks inhabited the waters and occasionally came within close proximity of the beach area.[57]

Occasionally, injured parties are able to recover against a government entity for failure to warn about a danger when the entity has specific knowledge that a danger exists in a specific place. In a 1951 federal court ruling, the federal government was liable when a grizzly bear injured a camper in Yellowstone National Park.[58] The visitor had asked a park ranger if it was safe to camp in a particular campsite. Although a bear had raided that particular campsite only a few days earlier, injuring several people, the ranger assured the visitor it was safe to camp there. Under the circumstances, the court ruled, the government had a duty to warn. In 1972, another bear attack at Yellowstone gave rise to a different federal court ruling.[59] In that case, the visitor had been given the usual warning brochures on the danger of bears, and park officials had no specific knowledge of bears in the area of the campsite where the attack occurred. Under these circumstances, the government was not liable.

Perhaps the most common fact pattern that produces litigation, with the government rarely found liable, are accidents involving deer hit by cars and trucks. Generally, governments have duties to post warning signs only when they have clear evidence that deer use a particular, defined area to cross a road with regularity. According to a New York ruling in 1986, government is liable only when the failure to erect a deer crossing sign would be irrational.[60] Generally, deer do not qualify as a "dangerous condition" under statutes that make highway departments liable for such conditions.

Even when government entities or their employees have been careless, injured parties can still fail to recover for injuries because the government activities are protected by sovereign immunity. A typical case is *Peterson v. Wyoming Game and Fish Commission,* a 1999 Wyoming decision involving a grizzly bear that mauled a hunter.[61] The particular bear was radio collared and part of a research project. It had been recaptured several times because it was killing livestock but had not been designated a nuisance bear so that it could be killed. The injured hunter alleged that the Wyoming Game and Fish Department was negligent in failing to control the bear. The court rejected the suit on the ground that sovereign immunity protected the agency and that none of the exceptions to immunity applied to the case. We might wonder whether the court would have ruled differently if the plaintiff was not a hunter.

Sovereign immunity sometimes bars suits also when visitors to zoos are injured by dangerous animals. At one time, government-run zoos were commonly found immune, but several states broke ranks and decided that zoos were a proprietary-type activity for which immunity was not appropriate. Today, the law varies considerably among the states. A 1996 decision found a Pennsylvania zoo immune when a dolphin bit off the finger of a volunteer.[62]

A 1978 Iowa decision found a city strictly liable for a tiger bite.[63] A 1970 Colorado decision held that a city could be liable, but only upon a showing of negligence.[64]

Perhaps the most interesting ruling in this area to be handed down in recent years came out of Arizona in 2004.[65] The fact pattern was rather typical, a motorist who collided with an elk. The ruling was unusual because the court implied that the state had a duty that went beyond merely putting up warning signs on the highway to mark high-accident areas. The signs, the state admitted, had no discernible effect in reducing accidents, as many studies have shown. In other places, the state had put up fences to keep elk from entering highways and, more importantly, had constructed underpasses to allow elk to get from one side of the highway to another. The underpasses, particularly in combination with fences, formed wildlife corridors that protected wildlife as well as motorists. According to the state game and fish department, the fence-underpass combination could reduce highway collisions by as much as 96 percent.

The ruling, *Booth v. State,* gives a glimpse of a legal regime that imposes greater duties on states to take care of wildlife. The court's concern was chiefly, if not solely, about motorists and their cars. Yet the wild animals involved in collisions are hardly disinterested bystanders. Animals are important public property that the state is supposed to protect under its wildlife trust duties. Perhaps in coming years, more courts will step forward and insist that states lay out wildlife corridors and otherwise consider measures to reduce human-animal conflicts.

CHAPTER
8
Indian Tribal Rights

The third group of governments in the United States are the many Indian tribes, which fit uneasily into the governance systems of the United States. Tribes possess distinct governance powers of their own, not derived from the federal or state governments. They also possess extensive property rights in land and natural resources arising from legal sources quite different from rights held by non-Indians. The powers of tribes over Indian reservations are reasonably well known. Less familiar and more contentious are the vested rights they possess to take fish and game at places outside their reservations. These off-reservation powers have given rise to recurrent legal disputes dealing with the precise scope of the tribal rights and with the complex ways tribal powers fit together with state powers, both on and off reservations and over tribal members as well as others.

To make sense of this important piece of wildlife law we need to begin with a primer on Indian law. What is the legal status of tribes, how do treaties fit into our system of laws, and what principles govern treaty interpretation? With these basics, we can take up the two central topics of the chapter: what wildlife-related property rights do tribes possess, and how do tribal powers over wildlife and wildlife harvesting fit together with state regulatory powers?

TRIBES AND TREATIES

Indian tribes occupy a legal status unlike any other entities in American law. They are not states, they are not foreign governments, nor are they like departments of the federal government. Their unique legal status has been the subject of various rulings by the U.S. Supreme Court, beginning in the nation's earliest years. As the Supreme Court has explained the roles of tribes, it has talked also about the treaties between tribes and the United States: about where treaties stand in the legal hierarchy and how courts should interpret them.

Tribal Status and Trust Obligations

The legal status of tribes was initially defined in 1831, when Chief Justice John Marshall took the position that they were neither states in the Union nor foreign governments.[1] Instead, tribes were "domestic dependent nations," a term Marshall coined to avoid the practical consequences of recognizing the inherent sovereign powers of tribes. According to Marshall's framework, tribes resided in a "state of pupilage" to the federal government, a relationship that Marshall likened to that of a ward to his guardian. As for internal matters, the tribes were self-governing entities, which meant that state laws had no application on tribal lands.

The independence of tribal reservations from state law, as we shall see, is no longer so complete. Legal relationships have grown more complex. The basic framework, though, remains as Marshall described it. Rulings in recent decades have described tribes as "unique aggregations possessing attributes of sovereignty over both their members and their territory."[2] The tribes "have been implicitly divested of their sovereignty in certain respects by virtue of their dependent status," but they nonetheless enjoy "an historic immunity from state and local control" while retaining "any aspect of their historic sovereignty not inconsistent with the overriding interests of the National Government."[3] Significantly, Indians have the right to make their own laws and be governed by them, a principle that requires "an accommodation between the interests of the Tribes and the Federal Government, on the one hand, and those of the State, on the other." Numerous rulings by the Supreme Court have fleshed out these broad principles.

Corresponding with this quasi-independent status of tribes is the duty of the federal government to exercise high standards of care in dealing with tribes and safeguarding their interests. The guardian-ward relationship, first expressed by John Marshall in 1831, has given rise over time to an extensive body of precedent. The federal government occupies the office of trustee, with tribes as beneficiaries. As trustee, the government owes fiduciary duties to deal with tribes fairly and to safeguard their interests—duties that sometimes obligate the federal government to bring lawsuits to safeguard tribal interests. The federal government cannot negotiate with Indian tribes aggressively at arm's length as it would other entities. On various occasions, courts have found that the federal government has breached these trustee duties.

Treaties and Reservations of Rights

As the United States pushed its boundaries westward, it entered into numerous agreements with Indian tribes. It did so in recognition that the tribes possessed legal rights in their lands that the federal government was bound to respect. According to Chief Justice Marshall, writing in 1835, the "hunting grounds" of tribes "were as much in their actual possession as the cleared fields of the whites; and their rights to its exclusive enjoyment in their own way and for their own purposes were as much respected, until they abandoned them, made a cession to the government, or an authorized sale to individuals."[4] Thus as tribes entered into treaties, they did not *acquire* property rights in land from the federal government. Instead, the tribes retained or *reserved* rights that they already possessed. Although the lands that they reserved became part of the United States and of the individual states, tribal property rights were not created under federal or state law.

Almost since the beginning, courts have been sensitive to the challenges tribes faced when negotiating with the United States. The treaties that memorialized these negotiations were written in English, and communication problems were vast. Tribes often experienced understandable difficulties traversing the legal and cultural gaps. In response to these problems and by way of leveling the playing field, courts have long interpreted treaties using canons of construction that strongly favor tribes. Repeatedly they have stated that treaties are liberally construed in favor of the tribes. Ambiguities are interpreted in their favor. In addition, courts interpret treaties as Indians at the time would have understood them, not according to the understandings of contemporary federal agents (and much less the interpretations readers today might give them).

These interpretive principles apply not just to ambiguities in what is written but to the larger transactions of which treaties were a part. For instance, courts interpreting treaties have considered the larger factual contexts of particular treaty negotiations—including the often-weak bargaining position of the tribes—to decide that treaties reserved rights to harvest wildlife, even though the reservations were not clearly expressed in the written document. They have given particular weight to the fact that hunting and fishing rights were so vital to many tribes that the tribes would never have signed a treaty relinquishing the rights. This reality has led courts to interpret treaties as reserving such harvesting rights even when not expressly mentioned.

Once treaties are interpreted to reserve particular tribal rights, courts are reluctant to conclude that later federal actions have curtailed or limited these rights or powers. Any federal action that might arguably abridge them is construed narrowly to minimize any reduction of tribal rights and autonomy.

Dissimilar Rights

As a consequence of these many treaties, members of Indian tribes typically have legal rights that non-Indians do not possess. As the Supreme Court has noted, however, treaty rights to hunt and fish are "servitudes"—the general term for property rights such as the hunting easements examined in chapter 4.[5] These rights thus are a type of property entitlement. They therefore have a sound legal basis and do not violate any constitutional principles requiring equal protection of laws.

The issue of tribal powers arises frequently in wildlife settings, especially when tribal members are able to hunt and fish free of limitations (including No Trespassing signs) that non-Indians must respect. Typical of judicial rulings considering the special status of tribes is one handed down in 2003 by the Supreme Court of Montana.[6] A non-Indian was charged with big-game hunting on tribal lands. State law banned non-Indians from hunting while allowing tribal members to hunt. The court rejected the defendant's claim that this arrangement violated the equal protection of laws. The statute was valid, the court held, because it was "rationally tied to the fulfillment of the unique obligation toward Indians."

Power of Congress to Terminate Rights

Treaties have no expiration dates and thus reserve rights for future Indian generations. (The treaties do expire, however, if the tribe goes out of existence as a recognized entity, even if descendants of tribal members remain.) Nonetheless, treaties are akin to other legal actions taken by Congress in that Congress retains the power to change its mind. Congress, in other words, can abrogate the terms of any treaty, with Indian tribes or with foreign governments, simply by enacting a statute to that effect. Indian treaty rights are thus precarious in a strictly legal sense: they can end at any time.

When the United States abrogates an international treaty, the abrogation ends both the commitments made by the United States to other countries and their corresponding commitments to the United States. The parties are returned to their positions before treaty signing. In the case of tribes, however, the situation is one-sided and, in an important sense, less fair. The federal government no longer has to live up to its obligations, but tribes do not thereby get their lands back.

Despite the manifest inequality of this arrangement, the Supreme Court has upheld the power of Congress to terminate treaty commitments. The Court has stipulated, however, that Congress can abrogate treaties completely or in part only if it does so expressly. Its intent must be clear and explicit. The Court explained this rule of law in a 1979 ruling involving Northwest fisheries: "Absent explicit statutory language, we have been extremely reluctant to find congressional abrogation of treaty rights. . . . Indian treaty rights are too fundamental to be easily cast aside."[7]

As a political matter, Congress has rarely seen fit to tamper with tribal rights, despite its power to do so. One situation in which Congress did exercise its power reached the Supreme Court in 1986.[8] The Yankton Sioux possessed an express treaty right to hunt bald and golden eagles on their reservation. In the Bald and Golden Eagle Protection Act of 1962, Congress severely restricted eagle hunting everywhere and explicitly required Indians to obtain permits when they sought to take eagles for religious purposes. The Court concluded that the 1962 act effectively terminated the Yankton's treaty right to take eagles. It held further that, due to the termination of the treaty right, the tribe enjoyed no exemption from the later-enacted Endangered Species Act of 1973.[9] The 1962 act, the Court explained, "reflected an unmistakable and explicit legislative policy choice that Indian hunting of the bald or golden eagle, except pursuant to permit, is inconsistent with the need to preserve those species." Using similar reasoning, a federal district court in 2000 ruled that Congress partially abrogated hunting rights of the Blackfeet tribe when it created Glacier National Park and banned all hunting within the park.[10]

Perhaps the most frequently heard claim concerning alleged abrogations of treaty rights is that Congress *implicitly* terminated rights when it enacted the various statutes admitting states to the Union. Congress undoubtedly possessed the power, when it enacted a statute admitting a new state, to terminate some or all of the tribal treaty rights existing within the state. Whether it did so when it admitted various states is a legal issue that continues to resurface, mostly because of a still-valid 1896 ruling by the Supreme Court, *Ward v. Race Horse.*[11] In that case, the Court held that, at the moment Congress admitted Wyoming as a state, it implicitly terminated rights of the Bannock tribe to continue hunting on all "unoccupied" lands owned by the federal government. The Court relied in its reasoning chiefly on the equal footing doctrine, the little-known but important provision of the Constitution that allows Congress to admit new states only if they are placed on an equal political footing with existing states. (For instance, a state cannot be admitted and allowed only one U.S. senator.) To admit Wyoming while the Bannock tribe retained extensive hunting rights, the Court reasoned in 1896, was to admit the state on unequal terms.

Hardly had the Court handed down *Race Horse* than it began expressing doubts about the reasoning and outcome. Within a decade it rejected a similar argument having to do with tribal fisheries in Washington State.[12] The Court has never expressly overruled *Race Horse,* presumably because doing so would upset land titles in Wyoming, a disruption the Court would prefer to avoid. For the past century, however, the Court has been hostile to claims that the admission of a state ended treaty rights. Most recently, in 1999, the Court expressly rejected *Race Horse*'s reasoning, making clear that "an Indian tribe's treaty rights to hunt, fish, and gather on state land are not irreconcilable with a State's sovereignty over the natural resources in the State."[13] As a precedent, in short, *Race Horse* seems to have little if any validity beyond the interpretation of the Bannock treaty in Wyoming. The nineteenth-century ruling, though, continues to arouse hopes among various non-Indian groups, who would like to see an end to off-reservation tribal rights.

The 1999 Supreme Court case, *Minnesota v. Mille Lacs Band of Chippewa Indians,* illustrates how treaty abrogation disputes can become complex and how they ultimately turn on the exact language and legal effect of a statute, presidential decree, or other legal action. A key 1837 treaty with the Chippewa reserved for the tribe extensive rights to hunt, fish, and gather wild rice on the lands they ceded under the treaty. These reserved tribal rights, however, lasted only "during the pleasure of the President of the United States." One legal claim in *Mille Lacs* was that these treaty rights ended when President Zachary Taylor signed an executive order in 1850 directing removal of the Indians. The president's order did contain a clear revocation of the treaty rights as part of an overall order for the tribe to move; no one disputed that fact. The Supreme Court concluded, however, that the president lacked the legal power to direct the Indians to move to a new location. Because the entire removal effort was invalid, the order terminating treaty rights was also invalid. The Chippewa thus retained their off-reservation rights. Even as it ruled this way, however, the Supreme Court emphasized that a president at any time could end the Chippewa's tribal rights simply with the stroke of a pen.

No Congress or president has shown much interest in terminating tribal treaty rights. The protections that tribes enjoy, though, are chiefly political and cultural, not legal. Tribes are thus wise to exercise caution as they assert off-reservation rights. A strong popular backlash among American citizens could change the political landscape. Congress or (in appropriate cases) the president could act at any time to end unpopular tribal rights.

THE SCOPE OF INDIAN PROPERTY RIGHTS

Indian tribes, as noted, possess vast power to control activities within the borders of their land reservations. This includes the power to control wildlife and wildlife harvesting. Tribal reservations are no longer absolutely immune from state regulation, a point taken up below. Nor are they exempt from all federal laws. Still, tribal powers on reservations are vast, including, at times, the authority to control and punish nontribal members.

Off-Reservation Rights

Less clear and more variable are the rights tribes have to engage in hunting, fishing, and related activities on lands they have otherwise relinquished in accordance with treaties. As noted, the Chippewa in Minnesota in 1837 reserved hunting, fishing, and rice-harvesting rights in the vast territories they were ceding to the United States under the treaty. Similar rights were reserved over Chippewa lands relinquished in Wisconsin. Best known, in terms of reserved tribal rights, are the rights that were reserved by numerous tribes in the Pacific Northwest. In 1854 and 1855, various tribes signed nearly identical treaties with the United States, all drafted and overseen by Isaac Stevens, superintendent of Indian Affairs and governor of the Washington Territory. These Stevens treaties (as they are termed) featured a common paragraph, repeated nearly verbatim, dealing with hunting and fishing rights. In terms of off-reservation rights, the treaties reserved "the right of taking fish at all usual and accustomed places, in common with citizens of the Territory, and of erecting temporary buildings for curing them; together with the privilege of hunting, gathering roots and berries, and pasturing their horses and cattle upon open and unclaimed land."[14]

The Stevens treaties have engendered major, highly contentious controversies, continuing to this day. We'll consider several chapters of that saga below. Here we need only note the overall content of the reserved tribal rights. The signing tribes retained rights to continue fishing at the "usual and accustomed" places that tribal members had long used, even though the lands were being ceded to the federal government. Their rights extended to the erection of temporary buildings to cure the fish. Moreover, the tribes retained rights to hunt, gather, and graze livestock on all "open and unclaimed land." These rights, too, remain valid and enforceable to this day. In combination, these treaty provisions give northwestern tribes extensive rights to harvest wildlife in places sometimes far removed from tribal reservations.

Because tribal off-reservation rights are based on treaty language—and because treaties vary greatly, despite the uniformity of treaties drafted by Isaac Stevens—it is difficult to generalize about the rights that tribes have. It is necessary to consult the specific language of each tribal treaty, and sometimes several different treaties, to determine the precise rights a tribe has reserved. Even then, treaties can make little sense without knowledge of special facts that are not set forth in the treaties. For instance, the various Stevens treaties reserve rights to fish at "usual and accustomed places" but do not enumerate where the places are located. External evidence, apart from the written treaty, is needed to locate these places.

Priority Over Private Property Claims

The most contentious issues relating to tribal off-reservation rights arise in situations where these rights cover lands that have passed from federal or state ownership into private hands. Private property rights in land originate in grants (that is, patents) issued by government. The private rights then pass by deed from owner to owner to the present. The question that people repeatedly ask is whether owners who buy land privately on the open market today take the land subject to these tribal rights—the "servitudes" noted above. And does the answer depend on whether the purchasers were aware of the tribal rights when they bought?

The priority of tribal rights has been litigated in various cases, and with uniform results. Under now-settled law, tribes hold their property rights based on aboriginal title, which predates all other property claims, private and public. Tribal rights, that is, are reserved by the tribes, not acquired by them in the treaties. Accordingly, when the federal government acquired tribal land by way of treaties, it took the ceded land subject to these outstanding property rights. Under basic property law, an owner of property cannot transfer to a later person more rights than the transferor possessed. Because it lacked full title to lands, the federal government could not and did not convey full title (that is, title free and clear of tribal claims) to anyone else. Thus land that passed to state ownership remained subject to these tribal rights. So did land that passed into private hands.

This issue of tribal rights in land now owned by non-Indians was raised in 1887 in an important ruling by the Supreme Court of Washington while it was still a territory.[15] A private landowner constructed a fence on his land that blocked access by Indians to one of their "usual and customary" fishing spots. The private owner claimed he had the right to fence the land as a matter of state property law. The court disagreed. The private owner, it ruled, acquired only the land title that the government originally held. That title was encumbered by the reserved tribal rights. This legal arrangement was confirmed in a key U.S. Supreme Court decision in 1905, *United States v. Winans.*[16] There the Supreme Court held that Indians held their off-reservation rights as servitudes on the land that were binding on all subsequent owners of that land. States were powerless to extinguish or curtail these rights.

Tribal Ownership of All Tribal Property

An important characteristic of these off-reservation rights—and of on-reservation rights as well—is that the tribes as legal entities possess the power to make binding agreements relating to the treaty rights. If it chooses, a tribe can enter into a treaty with the United States that alters or relinquishes the rights of all tribal members. Under the internal law of many tribes, these hunting and fishing rights are privately owned by particular Indian families and individuals, much as non-Indians own land under the laws of various states. Further, as a matter of internal tribal law, a tribal government might have no power to transfer or take action with respect to these individual private property rights. Nonetheless, U.S. law is clear in recognizing the power of tribes to act on behalf of all tribal members, without regard for tribal law and the internal limits on tribal powers.

This recognition of tribal power has significant implications. On one side, it vests tribal leaders and governing bodies with authority to make decisions on behalf of the tribe as a whole. It is not necessary for every tribal property holder to consent to an action that affects many landowners, just as it is unnecessary for every citizen of a state to assent to actions by the state. This tribal power makes matters far easier for the federal government, states, and even private parties that deal with tribes. If the tribe consents to a particular property transaction (to a mineral lease, for instance), then there is no need to get approval of each affected tribal property owner individually. On the other side, this rule of law diminishes the legal protections enjoyed by individual Indians owning tribal lands. The tribal government can sell their rights without asking permission. Further, the tribe may have no obligation to turn the proceeds of sale over to the individual owner.

This legal issue arose in a decision, *Whitefoot v. United States,* that reached the U.S. Court of Claims in 1961.[17] In 1954, the United States entered into an agreement with the Yakama Nation, under which the Yakama relinquished certain fishing rights along the Columbia River (for construction of the Dalles Dam) in exchange for a payment of $15 million. The relinquished fishing rights included the rights to fish at six "usual and accustomed fishing stations" that were owned personally by Minnie Whitefoot, a tribal member. The tribe received payment for these six stations but did not turn the money over to Ms. Whitefoot. Instead, it divided the money among all tribal members—$3,270 per member. Whitefoot sued the United States, seeking compensation for the loss of the stations. The court rejected her claim, upholding the power of the tribe to transfer the rights and to collect payment on Whitefoot's behalf. The court went further to assert (in a statement not legally relevant to the dispute) that Minnie Whitefoot also had no claim against the tribe itself based on the way it divided up the money.

A similar ruling was handed down by a federal appellate court in 1975, citing the *Whitefoot* decision.[18] The appellate court held that a tribe had the legal power to act on behalf of all tribal members with respect to tribal property, without regard for conflicting claims of individual property members. The court observed, though, that disputes among members of a tribe should be left to the tribe itself to adjust internally; it was inappropriate for U.S. courts, it asserted, to insist that internal tribal rights be based on principles of concurrent ownership incorporated into U.S. law. A better approach, the court explained in a footnote, was to allow the tribe itself "to arbitrate among the conflicting claims of its members according to the values and customs of their own culture."

STATE REGULATORY POWERS

In many settings, states desire to exercise control over the hunting and fishing activities of tribal members, particularly off reservations. Typically this regulation is advanced in the name of resource conservation. What powers do states have to regulate tribal harvesting activities ? The answer is complex because state powers vary a great deal with respect to the many tribes. Typically, tribes have the greatest immunity from state regulation when tribal reservation lands are owned entirely by tribal members. State power rises as more reservation land is owned by nonmembers. Also relevant is the fact that many tribes work out special arrangements with the federal government under which they exercise unusually broad legal authority over tribal lands, sometimes including authority over nontribal members living on the lands. As we have seen, federal law can preempt state law. Special wildlife management arrangements between tribes and the federal government can have the effect of preempting regulatory power that the state would otherwise possess. Despite this considerable tribe-by-tribe diversity, it remains possible to enumerate several generalizations about state regulatory powers. These generalizations supply a framework that can be filled in with details peculiar to a given tribe.

Tribal Members on Tribal Lands

For starters, states typically have little or no power to apply their fish and game laws to the activities of tribal members on reservations. The U.S. Supreme Court explained the typical arrangement in its 1983 ruling in the case of *New Mexico v. Mescalero Apache Tribe:*

> The sovereignty retained by tribes includes "the power of regulating their internal and social relations." A tribe's power to prescribe the conduct of tribal members has never been doubted, and our cases establish that "absent governing Acts of Congress," a State may not act in a manner that "infringed on the right of reservation Indians to make their own laws and be ruled by them."[19]

In this case, the state of New Mexico conceded that the Mescalero Apache tribe exercised exclusive jurisdiction over on-reservation hunting and fishing by tribal members. Its legal concession reflected the widely held view. Still, the Supreme Court has observed (in 2001) that "the Indians' right to make their own laws and be governed by them does not exclude all state regulatory authority on the reservation."[20] The Supreme Court's guidance has been scarce and vague. In *Mescalero,* the Court stated that "in exceptional circumstances a State may assert jurisdiction over the on-reservation activities of tribal members."[21] More recently (in 2001), it has announced that, when conservation or other "state interests outside the reservation are implicated, States may regulate the activities even of tribe members on tribal land."[22] States may have powers, for instance, to protect state-listed endangered species on tribal land or, in exceptional cases, to control hunting and fishing that threatens to destroy a wildlife population. Vagueness will likely linger until the Supreme Court has occasion to provide more precise guidance.

By and large, a state's powers to regulate tribal activities on tribal lands are unlikely to extend to the regulation of actual hunting and fishing activities. A Utah appellate court in 2006, for instance, struck down convictions of several Indians charged with hunting deer on a reservation without a license or tag.[23] The court held that the state could enforce its laws only if the tribe, in accordance with a state statute, agreed to allow the state to do so. Similarly, a Washington State appellate court in 2007 overturned a conviction of an Indian for unlawfully using gillnets to catch fish in tribal waters.[24] State law, the court announced, simply did not apply. Moreover, a state court did not have the power to convict the Indian even if the Indian's behavior violated tribal hunting laws.

This apparently exclusive tribal power on reservations does not undercut the ability of states to serve legal process on Indians on tribal lands. In an important 2001 ruling, *Nevada v. Hicks,*[25] the U.S. Supreme Court confirmed that states can enter tribal lands to serve papers and execute search warrants in the course of investigating offenses committed off the reservation. In addition, Congress has enacted a statute that authorizes certain states to apply their criminal laws on all or many tribal reservations within the state boundaries.[26] Significantly, this authorization expressly excludes hunting, trapping, and fishing laws, but does allow other laws to govern. Wisconsin is one state covered by the federal statute. In accordance with it, a Wisconsin appellate court in 2007 upheld the conviction of an Indian for hunting with a firearm on a reservation.[27] The Indian was a convicted felon, and state criminal law prohibited possession of firearms by a felon. The court decided that the ban on possessing the firearm was not a restriction on tribal hunting. Instead, it was a separate criminal law of statewide application, a part of the punishment that the defendant incurred due to his earlier conviction of a felony.

Nontribal Members on Tribal Lands

State power is more expansive when states seek to regulate the on-reservation activities of *nontribal* members. Tribes exercise broad control over their reserved lands, and thus over the activities of nontribal members who are on their lands. But tribal power is not as exclusive when the people affected are outsiders. In the case of some reservations, tribal power remains vast. For instance, in the 1983 ruling involving the Mescalero Apache tribe, the Supreme Court held that New Mexico had no authority to regulate on-reservation wildlife activities by nontribal members.[28] State regulation was entirely preempted by federal law. That dispute, however, involved unusual facts. In the peculiar case, the tribe had entered into a "sustained, cooperative effort" with the federal government to develop a detailed, professional fish and game management program. The federal-tribal program included fish hatcheries operated by the federal government to help sustain on-reservation fishing. The Supreme Court acknowledged that "under some circumstances a State may exercise concurrent jurisdiction over non-Indians acting on tribal reservations." But state authority was subject to being preempted by federal law, as it was in the instance of the Mescalero Apache Tribe.

The Supreme Court discussed at greater length the issue of state control over non-Indians in the 2001 case of *Nevada v. Hicks.*[29] That case involved the power of tribal courts to hear lawsuits brought against state officials for allegedly violating the rights of tribal members when they executed search warrants related to potential violations of federal law. In language that went well beyond what was necessary to resolve the dispute, the Court spoke generally to the powers of states on tribal reservations. "Our cases make clear," the Supreme Court stated, "that the Indians' right to make their own laws and to be governed by them does not exclude all state regulatory authority on the reservation. State sovereignty does not end at a reservation's border." State law, to be sure, was "generally inapplicable" when conduct on the reservation involved only Indians. But when conduct on a reservation involved non-Indians, the Court implied that state power was constrained only to the extent that it interfered with "what is necessary to protect tribal self-government or to control internal relations" within the tribe. Ownership of the land where an activity takes place is an important factor that affects the strength of state and tribal powers. State powers are less, and tribal powers correspondingly greater, when the nontribal activities take place on land owned by the tribe or by tribal members. State powers are more—if not even exclusive—when the actions by nontribal members occur on land owned in fee by non-Indians. We'll return to this issue at the end of the chapter when discussing tribal powers over non-Indians.

Off-Reservation Tribal Activities

Far more frequent than disputes arising on tribal reservations have been disputes about the ability of states to regulate off-reservation hunting and fishing activities by tribal members. When tribal members are *not* exercising treaty rights, of course, they are subject to state regulatory control to the same extent as all other citizens. The situation becomes complex only when tribal members are exercising reserved treaty rights, such as rights in the Stevens treaties involving fishing at all "usual and accustomed" fishing stations. What rights do states have to regulate in such situations?

The starting point, as always, is with the treaty language itself. Treaties have the status of federal law and thus preempt all conflicting state laws and regulations. So often have Stevens treaties led to litigation that most of the leading judicial rulings have dealt with them. The Columbia River salmon fisheries were historically almost unbelievably abundant: in 1911, for example, nearly fifty million tons of salmon were canned. Since then, the catch has steadily declined. As the catch dropped, the states began to restrict harvest. Although the states argued that their actions were intended to conserve dwindling stocks, the effect was to allocate fish to two groups, commercial and recreational fishers, both of which fished primarily off the coast of Oregon and Washington. The result of these policies was the imposition of increasingly stringent restrictions on upriver fishers, most notably tribal fishers fishing at the usual and accustomed fishing stations. Are these state conservation restrictions legitimate, given that Indian fishing can deplete fish stocks? That is the recurring question.

Off-reservation treaty rights, to reiterate, do not arise under state law. They are protected by treaties, which have the status of federal law and which therefore preempt inconsistent state laws. On the other hand, treaties do not reserve entire wildlife populations for tribes. Moreover, treaties do not prohibit states from conducting conservation programs to protect wild species. The inherent conflict thus is rather easy to see. States cannot interfere with treaty rights, but they do have legitimate interests in protecting wildlife and halting wildlife declines. States can, that is, regulate off-reservation tribal fishing and hunting in the name of conservation, but only within limits.

Important guidance on this issue was offered by the U.S. Supreme Court in a 1942 ruling, *Tulee v. Washington,* one of many disputes involving off-reservation fishing rights under the Stevens treaties.[30] Washington State took the view that it had full power to apply its fishing laws to tribal fishers since their rights to fish were simply "in common with" non-Indians. The "in common with" language, the state claimed, meant that tribal members had the same rights to fish as anyone else. The Supreme Court disagreed. To the contrary, the Court ruled, the state had no power to require tribal members to purchase state fishing licenses. As for other state fishing laws, states could enforce against tribal members only "such restrictions of a purely regulatory nature concerning the time and manner of fishing outside the reservation as are necessary for the conservation of fish."

So valuable and culturally charged are the northwestern fisheries that the Supreme Court has had numerous occasions since 1942 to flesh out the powers of states to regulate consistent with the various Stevens treaties. The Court's rulings, of course, rest upon the exact language of these treaties. Still, the kinds of clashes arising in the Pacific Northwest have arisen, or will one day arise, elsewhere. The basic underlying issue in time will come up everywhere: when is a state legitimately enacting laws on conservation, and when is it instead improperly interfering with tribal off-reservation rights?

In a highly visible 1968 ruling, the U.S. Supreme Court concluded that Washington State could properly regulate "the manner of fishing, the size of the take, the restriction of commercial fishing, . . . in the interest of conservation, provided the regulation meets appropriate standards and does not discriminate against Indians."[31] This language emboldened northwestern states to believe that they could regulate fisheries as they wished, in the name of conservation, so long as they did not overtly discriminate against Indians. It appeared to many readers of the Court's ruling that states could treat Indians just as they treated all other fishers, denying them any special status, so long as the states laws were aimed at conservation. In the instance of Washington State, it adopted a regulation that allowed only sport fishing for steelhead and prohibited all uses of nets by anyone. A commercial net fishery, the state decided, "would be inconsistent with the conservation of the steelhead." The effect was to halt nearly all tribal fishing because tribes used nets and did not have sportfishing boats and equipment.

A legal challenge to Washington's regulation returned to the Supreme Court in 1973. The Court struck the law down as inconsistent with the applicable treaty.[32] Because Indians only fished with nets, the effect of the regulation was to allocate the entire fish run to non-Indian sportfishers. In the Court's view, the state had to make an allocation of the scarce fish between Indian net fishers and non-Indian sportfishers; it could not force tribes to bear the full burden of curtailing their fishing to achieve conservation. The dispute thus returned once again to the lower federal courts, which ultimately decided to set aside 45 percent of the fishery for the tribes and 55 percent for non-Indian sportfishing, an allocation that the Supreme Court ultimately upheld in 1977.[33]

These rulings and others have highlighted the basic conflict in this area. States have legitimate interests in protecting and conserving wildlife. Wildlife populations, in turn, transcend tribal and other boundaries. But how can a state conserve its wildlife if significant harvesting (including commercial harvesting) takes place outside state control—that is, by tribes who escape state regulation? Granted that tribal members have harvesting rights, what about all other citizens? On the other side, state efforts undertaken for conservation can effectively deprive tribes of their treaty rights. Regulations labeled as "conservation" often have allocation impacts. Merely labeling a regulation as a conservation measure does not keep it from being struck down when its impacts fall disproportionately on tribal members, as in the instance of Washington State's rule restricting steelhead fishing. As a practical matter, conservation requires some type of restrictions on harvesting. Restrictions, though, necessarily have the effect of allocating a scarce resource among users. In short, the function of resource conservation is not distinct from the function of resource allocation.

Courts have sought to reconcile this conservation-allocation conflict by insisting, first, that state regulations be truly necessary for conservation and, second, that they not discriminate against Indians. In the end, though, these two principles have not been enough to resolve disputes. It has been necessary for courts, or other lawmakers, to make express allocations of a resource between Indians and non-Indians, even allocations that seem arbitrary (for instance, Indians get half and non-Indians get half). No other approach seems to ensure that Indian rights are protected. In a 1979 ruling, the U.S. Supreme Court ruled that a harvesting allocation should begin by dividing a fish run "into approximately equal treaty and nontreaty shares," with the tribal share then reduced if tribal needs can be satisfied with a lesser catch (a caveat, it should be noted, that many Indian scholars perceive as biased against allowing Indians to fish commercially).[34] Typically, allocations exclude fish captured by Indians for ceremonial and subsistence purposes; these do not count toward the Indian's fish quota.

The treaty rights fishing cases have been, in effect, the civil rights cases of the Pacific Northwest. In its recalcitrance and unwillingness to recognize the rights of Indians, Washington State in particular has resembled the states of the American South during the civil rights era of the 1950s and 1960s, a resemblance that the federal appellate court in the region specifically noted.

Stevens treaty litigation has gone beyond allocation issues to address two further points having to do with state power. States sometimes revive depleted fisheries by operating fish hatcheries and releasing fry into waterways. The rule that seems to prevail is that Indian rights extend to a share of these hatchery-raised fish, even though states have no obligation to operate the hatcheries.[35] One reason for this ruling is that fish released into the rivers replace fish that non-Indian activities have eliminated to the harm of the tribes.

The more difficult issue, far from ultimate resolution, is whether tribes have the legal power to challenge activities that disrupt their fisheries in a physical way, as by polluting waterways, blocking fish migrations, and degrading waterways ecologically. What good is a right to fish if no fish are available to catch due to habitat loss? What if water diversions are so substantial as to cut fish populations severely? And if habitat changes are not reversible—as in the instance of massive dams that cannot be fitted with fish ladders—should tribes be able to claim substitute fishing rights in other locations to compensate for interferences with their treaty rights?

Few judicial rulings have provided much guidance on this question. A 2007 ruling by a federal court in Washington—continuing the long-running saga of litigation over Indian salmon fishing—considered whether Washington State could be compelled to reconstruct various road culverts on state roads to avoid blocking salmon migrations in salmon streams.[36] The court ruled that the "right of taking fish, secured to the Tribes in the Stevens Treaties . . . imposes a duty upon the State to refrain from building or operating culverts under State-maintained roads that hinder fish passage and thereby diminish the number of fish that would otherwise be available for Tribal harvest." If approved on appeal, this ruling could set the stage for a wide array of further lawsuits challenging activities that interfere with tribal fishing, at least (as the court explained) when the interference hampers the ability of Indians to make a "moderate living."

TRIBAL REGULATORY POWERS

The many legal issues about state power to regulate fish and game are matched by similar questions about tribal powers. What regulatory powers do tribes have, on and off the reservation and over tribal and nontribal members? The larger subject, again, is quite complex because much depends on the language of individual treaties. Further complications arise because lands on many reservations lands are fragmented between tribal and nontribal owners. Indeed, in some instances, nearly all reservation land has passed out of Indian hands.

As noted, tribes have vast powers to regulate the activities of their members within the confines of reservations. They can and sometimes do craft extensive regulatory codes governing hunting, fishing, and wildlife management. Tribes also typically possess the power to limit hunting and fishing on their lands by nontribal members. This latter power, though, is usually concurrent with state regulations. States can insist that nontribal members obtain fishing licenses and perhaps comply with other state laws as well, even if the laws do not govern tribal members. An exception to this arrangement was described in the 1983 Supreme Court ruling dealing with the Mescalero Apache tribe, mentioned earlier in this chapter. On the specific facts of the case, particularly the tribe's extensive cooperative wildlife management program, the Court ruled that state regulation was entirely preempted on the reservation, even over nontribal members.

In an important 1981 ruling, *Montana v. United States,* the Supreme Court considered whether a tribe could regulate hunting and fishing by nontribal members in a river, the bed of which was owned by the state, not by the tribe or tribal members.[37] The general rule, the Court stated, was that tribal powers extend only so far as necessary to protect tribal self-government and to control internal relations. The Court held that the tribe could not regulate the hunting and fishing—even though the state land fell within the outer boundaries of the tribe's reservation—without a showing that non-Indian hunting and fishing threatened the tribe's political or economic security. For a time, it seemed that this principle, announced by the Supreme Court, applied only when nontribal activities were taking place on nontribal lands. But in its 2001 ruling *Nevada v. Hicks,* the Court stated otherwise. In *Hicks,* the Court announced that land ownership was not an all-important factor. Who owned the land—Indians or non-Indians—was "only one factor to consider in determining whether regulation of the activities of nonmembers is 'necessary to protect tribal self-government or to control internal relations.'" The implications of this judicial language are by no means clear and will no doubt take time to emerge. Tribal powers over nonmembers are certainly constrained—that much is clear. *Hicks,* though, did not deal with wildlife on tribally owned lands—that is, with a resource that was essentially owned by the tribe itself. Tribal governance might always be deemed "necessary" to tribal self-government, given the vast importance of hunting and fishing to many tribes.

This brings us, finally, to the last issue of governing power, the power of tribes to regulate activities of tribal members when the members are exercising off-reservation hunting and fishing rights. A rare, leading ruling on this issue was handed down by a federal court in 1974.[38] The case dealt with alleged violations of Yakama Nation fishing regulations that applied off-reservation at the tribe's "usual and accustomed" fishing stations. The court concluded that the Yakama Nation, in its 1855 treaty with the United States, reserved not just the right to fish at the fishing stations, but the power to regulate tribal fishing. This power included an ability for tribal game officers to arrest tribal members caught in the act of violating the regulations at the fishing stations. The court supported its conclusion by referring to the *Whitefoot* decision, considered above, in which a court decided that tribal fishing rights at the fishing stations were owned by the tribes, not by individual tribal members. The determination of how fishing would be conducted at these stations was an "internal affair" of the tribe, and thus an appropriate subject for tribal regulation. As for the powers of tribal game officers off the reservation, they were clearly limited:

> Our holding that the Yakima Indian Nation may enforce its fishing regulations by making arrests and seizures off the reservation is a very narrow one. Off-reservation enforcement is limited strictly to violations of tribal fishing regulations. The arrest and seizure of fishing gear must be made at "usual and accustomed places" of fishing, and only when violations are committed in the presence of the arresting officer. Tribal officers patrolling off-reservation sites are subject to all reasonable regulations that may be imposed by the State of Washington for the orderly conduct of inspects, arrests and seizures.

The powers of tribes to engage in regulatory actions of all types are subject to alteration by Congress and, to a lesser extent, by federal administrative actions. These decisions, in turn, are subject to political pressures. So, too, are decisions by the Supreme Court influenced in part by the judgments of the justices. At this writing (2008), Supreme Court rulings display an inclination to view tribal powers with suspicion and to prune those powers. Wildlife, though, is an area in which courts for generations have shown special solicitude for tribal sovereignty. For a full century now, the Supreme Court has stuck closely to the judgment it expressed in 1905: the ability to hunt and fish was "not much less necessary to the existence of the Indians than the atmosphere they breathed."[39] Despite the leanings of particular justices, this judgment is likely to remain the starting point for future legal disputes.

CHAPTER

9

Key Federal Statutes

Wildlife law for the most part originates at the state level of government. The states own the wildlife within their borders as trustees and, as such, bear duties to care for wildlife in the public interest. For its part, the federal government has largely let them do so. The few federal statutes tend to address interstate issues that individual states find difficult to handle. The federal statutes also aid state conservation efforts by turning certain violations of state law into federal offenses so that federal investigators and prosecutors can assist in law enforcement efforts. Only in the case of a few categories of wild species does federal law engage in direct conservation. Beyond that (and as the following chapters explain), federal law plays significant roles in conserving threatened and endangered species and wildlife on federal lands.

This chapter examines three federal statutes: the oldest and most widely applicable federal wildlife law, the Lacey Act; the Migratory Bird Treaty Act; and the Bald and Golden Eagle Protection Act. We do not consider several more narrowly focused statutes, such as those that manage marine species (most notably the Marine Mammal Protection Act and the Magnuson-Stevens Fishery Conservation and Management Act); those that protect feral species (the Wild Free-Roaming Horses and Burros Act); and those that protect exotic animals imported from other countries.

THE LACEY ACT

The federal government got involved in wildlife conservation nationally in 1900 when it enacted the Lacey Act, a statute aimed at the nagging problem of game law violators who conducted business across state lines, frustrating state enforcement efforts. Since the frustration was largely traceable to the then-current understanding of the relationships between the federal and state governments, congressional action was necessary. As originally written, the statute did two things. First, it required people transporting wild animals or animal parts across state lines to label their packages clearly so that law enforcement officers could determine what they contained. Second, the act reinforced state laws by prohibiting the transportation of wild animals or animal parts that had been taken or possessed in violation of state law. Over the next eight decades, the statute was broadened to encompass a wider variety of wild species and a greater variety of laws, including Indian tribal laws, foreign laws, and international treaties, in addition to violations of state laws. The statute's scope gained breadth, but prosecutions under it remained few. Maximum penalties were relatively low. Also, felony prosecutions were difficult because of the heavy burden imposed on prosecutors to prove the defendant's actual knowledge and specific intent in order to gain a conviction.

The Lacey Act was extensively revised in 1981. Congress clarified and reorganized the statute's provisions, increased the maximum penalties, and reduced what prosecutors had to prove. These changes made the act a much-stronger conservation tool. Prosecutions have become more common.

The central element of the Lacey Act is section 3372(a), which prohibits anyone from handling wildlife in various ways if the wildlife was "taken, possessed, transported, or sold in violation" of a variety of wildlife laws or regulations.[1] A violation of the Lacey Act, that is, includes two elements. First, there must be an initial violation of some other wildlife law (the "predicate offense"). Then, the wildlife involved, dead or alive and in whole or in part, must be handled in one of the ways specifically listed in the statute—for example, by importing or transporting the wildlife.

As for the initial violation of another wildlife law, that law can be state law, Indian tribal law, foreign law, a federal treaty, or even another federal statute. Without the violation of some other law, however, there is no violation of the Lacey Act. (As noted below, this is not true in the case of the Lacey

Act's rules on marking wildlife and filling out wildlife declarations; also, since 2003, the Lacey Act has directly protected certain big cats.)

If the underlying statute or regulation that has been violated is either a *federal law* or an *Indian tribal law,* then any person who imports, exports, sells, receives, acquires, or purchases the wildlife violates the statute. If, instead, the underlying statute or regulation is either a *state law* or a *foreign law,* then the Lacey Act is violated when such a person engages in one of the same actions "in interstate or foreign commerce." Thus violations of the Lacey Act are somewhat harder to prove when the underlying law that has been violated is a state or foreign law. In those instances, the person accused of violating the Lacey Act must "import, export, transport, sell, receive, acquire or purchase" the wildlife in a transaction that is somehow linked to commerce among the states or among nations, which generally requires that the person or the wildlife crossed a state or international boundary.

The Lacey Act, like most modern federal statutes, contains its own definitions of the terms used. Two of the definitions are particularly important. The act applies to all fish or wildlife. That category includes all wild animals, dead or alive, "whether or not bred, hatched, or born in captivity." It also includes "any part, product, egg, or offspring thereof." The act therefore applies to captive animals, not just those taken from the wild. Thus a 2006 federal court ruling from Oklahoma, *United States v. Condict,* applied the statute to "farm raised domesticated deer."[2] The definition of "transport" is also significant because it includes not just moving a wild animal, but delivering or receiving it "for the purpose of movement, conveyance, carriage, or shipment." A person or business can therefore violate the act merely by turning wildlife over to a shipping agent or by receiving an item for shipment or other transport, without ever personally moving the wild animal any distance. Transport, in short, can take place without any motion.

Recent prosecutions illustrate how the two parts of a Lacey Act violation fit together. In a 2004 prosecution in Nebraska, a hunter was found guilty of shooting and killing a deer (two separate violations) in a closed portion of a national wildlife refuge, in violation of the federal law governing the refuge.[3] The hunter then removed the deer from the refuge and took it home. The violation of the refuge rules provided the underlying illegal act. The removal from the refuge satisfied the requirement of transport. Since the predicate violation was of a federal law, there was no need to prove that the transport crossed a state or international boundary.

A similar federal court case from 2001 involved a hunter who killed an elk on Ute tribal lands.[4] The taking of the elk violated several tribal laws—it was out of season and after dark (hunting was allowed only in daylight); the hunter had no permit; the hunter was not a tribal member; and the elk was not tagged. Again, by taking the animal home to eat it, the hunter engaged in transport. As with violations of federal law, violations of tribal laws do not require that the transport cross state or national boundaries.

In contrast to these cases is a 1991 ruling from the federal court of appeals in California, *United States v. Carpenter.*[5] A landowner who raised goldfish in an outdoor pond violated a federal statute, the Migratory Bird Treaty Act of 1918,[6] by hiring staff members to shoot thousands of herons, egrets, and other birds that were preying on the goldfish. The court agreed that the landowner violated the Migratory Bird Treaty Act but held that there was no violation of the Lacey Act because no action was taken with respect to the birds after they were killed. Without more, an action that violated the Migratory Bird Treaty Act did not also violate the Lacey Act.

Lacey Act violations based on underlying state laws formed the basis of a 2004 federal appellate ruling from Tennessee, *United States v. Hale.*[7] A Tennessee couple operated a caviar business using the roe of paddlefish. They routinely and knowingly bought the roe from fishermen who had taken paddlefish out of season. They then sold the caviar in Kentucky as well as Tennessee. Their Lacey Act convictions were upheld because the caviar was transported into Kentucky. Similarly, in a 2001 Oklahoma case, a federal appellate court upheld the conviction of a man who lured elk from a national wildlife refuge onto his own fenced land and then advertised in Texas for hunters to shoot them.[8] A hunter did so (in a penned area of only fifteen acres, where the "hunt" took ten minutes) and then returned to Texas with the carcass. The landowner violated Oklahoma law by failing to obtain various permits. The interstate advertising, along with the sale of the "hunting" experience, provided an adequate connection to interstate commerce. In its ruling, the court made clear that the state law being violated did not have to be a wildlife conservation law; it could be a revenue-generating licensing law.

A 1986 case involved a North Carolina resident who purchased rockfish that had been taken illegally in Virginia.[9] The man intended to market the fish in New York or Maryland, and sold it to a North Carolina company that he knew would transport the fish to those markets. His conviction was upheld because he "knew that the rockfish would be shipped in interstate commerce and he took steps that began their travel to interstate markets."

If reported appellate decisions are any guide, Lacey Act violations increasingly involve cases in which the underlying illegality is a violation of foreign law rather than a law from within the United States. These cases can sometimes become quite complex and raise difficult issues of fairness. An easy case arose in 2007 when Robert Kern, individually and through his company, arranged trips to Russia in which clients could shoot moose and sheep from helicopters in violation of Russian law.[10] The hunters brought their illegally obtained animal trophies back to the United States. The importation of the animals and the interstate commerce were clear, and Kern was convicted. A more complex case arose in 2003 when a Virginia company that sold primates for medical research contracted to buy fourteen hundred monkeys from an Indonesian company.[11] The monkeys included both captive-raised and wild ones. The export of the wild monkeys violated Indonesian law, but, as a result of bribes, Indonesian officials signed export documents. The court held that the export of the monkeys was unlawful despite the official export documents, and that a court in the United States did not intrude upon Indonesian sovereignty by challenging the validity of an action by an Indonesian official under Indonesian law.

Two other cases further illustrate the challenges and possible unfairness of applying the Lacey Act strictly to all types of foreign law violations. In a 2003 ruling, an importer of Caribbean spiny lobsters from Honduras was found guilty based on an underlying violation of Honduran law, even though Honduran officials subsequently declared the law invalid.[12] The U.S. federal district court had undertaken its own determination as to the validity of the Honduran law. The appellate court upheld the convictions on the ground that the Honduran law was in effect in Honduras at the time; its later invalidation by Honduran officials, even though intended to be retroactive, did not upset the Lacey Act conviction. In a 1991 case, various fishermen on a Taiwanese fishing boat were found guilty of Lacey Act violations when their boat conveyed salmon across the Pacific and then transferred the salmon, on the high seas, to an American vessel for import into the United States.[13] The salmon had been caught off the coast of Taiwan by a squid fishing vessel. A Taiwanese regulation prohibited salmon fishing by squid vessels. The regulation carried no criminal penalties, and only the captain of the squid vessel was subject to any sanctions, not the individual fishermen. Nonetheless, the federal appellate court upheld the fishermen's convictions. It made no difference that the foreign law was only a regulation, not a statute; that it carried no criminal sanction; and that only ship captains could violate it.

An additional category of Lacey Act cases merits special attention—cases involving people who provide outfitting or guiding services for hunters who subsequently violate state laws. The application of the Lacey Act to this group was somewhat vague until Congress revised the statute in 1988. As revised, the statute expressly states that an outfitter or guide engages in a "sale" of wildlife by offering or providing paid services, or by obtaining a hunting or fishing license for a customer that results in the "illegal taking, acquiring, receiving, transporting or possessing of fish or wildlife." Similarly, a customer of an outfitter or guide engages in a "purchase" of wildlife if the services or license is for an illegal act. The effect of these amendments was to make outfitters and guides liable for the wrongful acts of their customers, at least (in the case of state law violations) when either the customer is from a different state or animals parts are transported out of state. It makes no difference that the outfitter or guide offered the services and received payment before the illegal hunting activity took place and without a plan or arrangement for it to occur. Thus a Colorado hunting outfitter was convicted in 2003 when an unlicensed client killed a mule deer and the outfitter attached a tag that belonged to another hunter.[14] The outfitter also helped the same client kill a mountain lion, and then reported that he, the outfitter, had killed it. The client took both trophies home to Wyoming.

The Lacey Act contains a variety of penalties for violation.[15] Generally, the more severe the penalty, the greater the burden on the prosecution to show that the defendant knew specifically what was going on. *Civil penalties,* for example, can be imposed if the violator "in the exercise of due care should know" that the wildlife involved is illegal under "any underlying law, treaty, or regulation." If the market value of the fish and wildlife is less than $350 and the Lacey Act violation involves only transportation, acquisition, or receipt of the illegal wildlife (not purchase, sale, import, or export), then the civil penalty is the lesser of $10,000 or the maximum penalty imposed for the underlying illegal act. *Misdemeanor penalties,* on the other hand, require a showing that the defendant knowingly engaged in the conduct prohibited by the act (the import, export, sale, purchase, transport, and so forth) and either knew of the underlying illegality or "in the exercise of due care should" have known about it. Misdemeanor violations can include fines up to $10,000 and imprisonment up to one year.

The highest level of proof is needed to sustain a felony violation, and the definition of the offense is somewhat narrower. As for the offense, *felony penalties* cannot be imposed merely for transporting illegal wildlife. The prosecutor must show that the defendant knowingly engaged in either a purchase or sale of illegal wildlife or its transportation across state or international boundaries. Thus in Minnesota in 2005, one defendant avoided liability under the Lacey Act for transporting endangered species that had been the subject of unlawful trafficking because there was no evidence that the defendant had done more than carry them.[16] In contrast, a federal appellate court, in *United States v. Senchenko,* upheld the felony conviction of a bear hunter who collected too many bear gall bladders for personal use; the court held that the quantity of the bladders alone provided evidence of intent to engage in commercial activity.[17]

In addition, the prosecutor in a Lacey Act felony case must prove that the defendant knows of the underlying illegality (the predicate offense). It is not enough to show that the defendant, if reasonably careful, should have known of the illegality. This requirement of actual knowledge of the illegality has fueled much litigation. In one case from 2002, *United States v. LeVeque,* a guide escaped liability based on a customer's taking of a bull elk that was slightly too small to be legal in a locale designated as a "brow tine bull only" hunting area.[18] The court agreed that the guide should have known about the designation and rule in the exercise of due care. Although that knowledge was enough to satisfy a misdemeanor prosecution, the guide's lack of *actual* knowledge meant that the felony prosecution failed. In a 1988 case involving the sale of lobster tails that were too short to sell under state law, the prosecutor showed the defendant's actual knowledge of the illegality by introducing evidence of a prior conviction for violating the same law.[19] On the other hand, it is not essential for the prosecutor to prove that the defendant knew *what* law was actually violated. It is enough if the defendant knew the wildlife was in some way illegal. Thus in 2001, a federal court of appeals upheld the convictions of the owner of a tropical fish store in Southern California who had been found guilty of a felony for importing ten baby parrots from Mexico without declaring them at the border, because he realized that he was in some way possessing them illegally.[20] As a federal appellate court explained in 2004 in *United States v. Hale,* discussed earlier in this chapter: "the government's burden is to show that a defendant had knowledge that his or her conduct was illegal, not actual knowledge of the exact underlying statutes that were violated."[21]

Three further aspects of the Lacey Act will conclude our discussion of the act.

First, the Lacey Act from its beginning in 1900 has required packages containing animal parts shipped in interstate commerce to be clearly marked for ready identification. Marking rules are now promulgated as administrative regulations. Failures to label packages properly carry a maximum civil penalty of only $250 and have no criminal sanctions. Significantly more severe penalties are imposed for violating the Lacey Act's provisions banning false shipping documents or labels for any wildlife items coming from a foreign country or transported in interstate or foreign commerce.

Second, the act includes strict forfeiture provisions. Illegal wildlife items can be forfeited based on any violation, without a showing of knowledge or intent—there is no "innocent owner" defense. Thus in a 2005 ruling, a federal appellate court in Washington upheld the forfeiture of 144,774 pounds of blue king crab that had been taken in violation of Russian fishing regulations.[22] The importer's ignorance of the illegality was irrelevant. Generally (and subject to more complex rules), an owner can forfeit vehicles, aircraft, vessels, and other equipment used in illegal activities if the items were used in a felony and the owner of them (who may have had nothing to do with the felony) knew or should have known that they would be illegally used.

Finally, Congress in 2003 revised the Lacey Act in a minor but quite surprising way. Breaking with the act's one-hundred-year history, Congress expanded the statute to make it unlawful to "import, explore, transport, sell, receive, acquire or purchase in interstate or foreign commerce" any "live species of lion, tiger, leopard, cheetah, jaguar, or cougar or any hybrid of such a species" except in narrowly tailored circumstances.[23] The new prohibition applies only to live animals, not animal parts, and of course applies to very few animals in the United States. What is significant is that the prohibition applies whether or not the conduct covered violates any other law—state, tribal, foreign, or federal. For the first time (setting to one side the marking and false-labeling rules), Congress has expanded the Lacey Act to reach beyond its traditional role of supplementing other wildlife laws to ban undesired activities directly.

THE MIGRATORY BIRD TREATY ACT

Nearly a century ago, the federal government stepped forward to address declining populations of birds, a problem that states individually were unwilling or unable to handle. Migratory waterfowl were a particular challenge for individual states because protective measures taken in one state could simply provide more birds for hunters elsewhere to harvest. Collective action was needed. Birds were also at risk because of the demand for bird plumes by the clothing trade, especially in women's hats.

After an initial federal statute was struck down in court as beyond Congress's power, the federal government in 1916 entered into a treaty with Great Britain—which still handled international affairs for Canada—governing migratory birds in the United States and Canada.[24] In time, the 1916 treaty served as a model for three further agreements executed in following decades—a 1936 treaty with Mexico, a 1972 agreement with Japan, and a 1976 treaty with the Union of Soviet Socialist Republics.[25] Although the provisions of the treaties differ in minor ways, they have collectively committed the United States to protecting the vast majority of all migratory bird species within its borders. Congress implemented the initial treaty with the Migratory Bird Treaty Act of 1918.[26] The act has been amended several times to take into account later treaties, to enhance penalties for violations, and to address a variety of issues that have arisen over time.

The Migratory Bird Treaty Act (MBTA) has an unusual form. It begins by banning all actions that kill or harm migratory birds, and then authorizes the secretary of the interior to issue regulations that permit hunting, wildlife control, and other measures. The act's initial prohibition is as broadly (and redundantly) phrased as the statute's drafters could make it. It is "unlawful at any time, by any means or in any manner, to pursue, hunt, take, capture, kill, attempt to take, capture, or kill, possess, offer for sale, sell, offer to barter, barter, offer to purchase, purchase, deliver for export, ship, export, import, cause to be shipped, exported, or imported, deliver for transportation, transport or cause to be transported, carry or cause to be carried, or receive for shipment, transportation, carriage, or export, any migratory bird, any part, nest, or egg of any such bird, or any product, whether or not manufactured, which consists, or is composed in whole or in part, of any such bird of any part, nest, or egg thereof" if the bird species is encompassed by any of the four international treaties.[27]

On its face, this prohibition imposes strict liability; that is, a violation occurs without regard for whether a person intended to engage in the prohibited act or realized that the violation was taking place. As noted below, however, serious penalties are imposed only upon a showing of some level of knowledge or intent.

The bird species covered by the Migratory Bird Treaty Act is far broader than most people would guess. Many bird species are fully migratory in the sense that all individual birds move seasonally from one location to another, usually between breeding grounds and wintering grounds. Other bird species inhabit a particular territory with little movement, or they inhabit wide territories but retreat from some of the territory during harsh weather. These differences are irrelevant; all are covered. In fact, as a legal matter, a bird is migratory if it is so designated by the U.S. Fish and Wildlife Service in a list published in the *Federal Register* and the *Code of Federal Regulations.* The most recent list, published on August 24, 2006, although not yet official, includes the vast majority of all bird species, whether or not they display significant migratory patterns—some 972 bird species in all. The list publication is an agency action that can be challenged in court when it takes place. Once the list goes into legal effect, however, a person accused of violating the act cannot then claim that a species was improperly included. In a 1989 federal district court case from Kansas, for instance, a landowner was found guilty of killing great horned owls that preyed upon his chickens.[28] Because the owl was included on the official list, the court refused to consider the claim the owls were year-round residents of the region and didn't actually migrate, even though the assertion was factually accurate.

As discussed in chapter 1, a recurring issue under the Migratory Bird Treaty Act is its application to birds raised in captivity. The original statute recognized the problem. It provided that nothing in the act should be interpreted "to prevent the breeding of migratory game birds on farms and preserves and the supply of birds so bred under proper regulation for the purpose of increasing the food supply."[29] Regulations issued by the Fish and Wildlife Service authorize the raising, transporting, and processing of mallard ducks (the most common species raised in captivity) and other migratory birds. Sometimes permits are required. In most circumstances, parties only need to mark the birds prominently in specified ways. As a federal appellate court ruled in 2006, states can supplement these rules with their own permitting requirements.[30]

The problem arises when captive-bred birds are not sold directly as food but instead are released for hunters. Over the years, a few courts have concluded that the Migratory Bird Treaty Act applies only to wild animals, not to captive-bred ones, usually citing the language in section 711. The dominant sentiment, however, has been otherwise, and the statute itself makes no such distinction. (It applies to all migratory birds, and migratory birds are all birds on the species list issued by the Fish and Wildlife Service. The applicable regulations make the point abundantly clear.) The complexity of the issue was highlighted in a 1961 federal appellate ruling involving a landowner in Minnesota who ran a game ranch, setting out bait and charging hunters to enter to shoot ducks.[31] The landowner raised mallards in captivity and released them on his ponds. The captive-raised ducks, however, were physically indistinguishable from the wild ducks and intermingled with them. Many of the ducks migrated over the winter, and some returned the following year, but the landowner had no way of knowing if any of the returning birds were captive bred. Although the court assumed that the statute applied only to wild ducks, not to tamed or domesticated ducks, that simply raised a second issue: which ducks were wild under the circumstances? To answer, the court looked to the common law rules on property rights in animals (considered in chapter 3). It concluded that, once the landowner released the ducks onto his ponds, they had gained their liberty, were no longer the landowner's property, and thus were wild birds protected by the statute.

Current regulations clarify this problem of intermingled wild and captive-raised birds. (They also state clearly that the act applies to hybrids of listed species.) In general, owners can kill captive-raised birds in any way and in any number so long as the birds are properly marked. Shooting, however, is controlled. Generally, captive birds can be shot only under the same rules on seasons, bag limits, duck stamps, and licensing requirements as applied to wild birds. An exception exists for mallards (but not other waterfowl species) that are shot "within the confines of any premises operated as a shooting preserve under State license, permit or authorization."[32] Even on shooting preserves, the ducks must be specially marked. A related exception includes mallards used in dog training trials.

A second issue, also having to do with the species covered by the Migratory Bird Treaty Act, is its application to migratory birds that are present in the United States only as a result of human efforts, intentional or unintentional. The treaty with Britain on behalf of Canada lists the protected species by family and genus, not by individual species. The listed families include many species not native to North America, including many introduced species. The issue came to a head in a 2001 ruling by the federal appellate court in Washington, D.C., in a dispute related to the mute swan, which is not a native species.[33] The swan belongs to the family Anatidae, which was included as a family in the treaty with Canada. The U.S. Fish and Wildlife Service, however, omitted the particular species from the migratory bird list, mostly due to its aggressive territoriality and its tendency to displace other species, including threatened and endangered species. The federal court ruled that the omission of the species was arbitrary and capricious, given the clarity of the treaty language. The Fish and Wildlife Service then began protecting the bird, although without formally listing it.

Congress responded in 2004 by amending the Migratory Bird Treaty Act to exclude all species introduced by humans to the United States, including the mute swan.[34] The sole exception was for species that were native and present in the United States, then disappeared, then were reintroduced as part of a federal program. The complication of this statutory change was that it conflicted with express treaty language that extended protection to entire families of birds, without reference to whether the birds were native. The organization seeking protection for mute swans promptly challenged the regulations implementing the new statute, claiming that they were invalid because of a conflict with the underlying treaty.

When it revised the statute to exclude nonnative species, Congress included a provision expressing, as "the sense of Congress," that the amendment was consistent with the four treaties. Although this language had no legal effect, it was added presumably to protect the amendment from legal challenge. The federal court hearing the new challenge upheld the regulations, including the exclusion of the mute swan.[35] It did so, ironically, by ignoring the "sense of Congress" language. The court concluded that, despite the statement by Congress, the new statute was in fact in conflict with the treaties and that Congress in effect intended to abrogate the treaties in part—a power that it undoubtedly possessed. In other words, the statute was valid only because Congress made abundantly clear that it wanted to exclude introduced species and because a later act of Congress (the 2004 amendment) took precedent over earlier acts (approving the four treaties).

On March 15, 2005, the Fish and Wildlife Service published its list of introduced species not protected by the Migratory Bird Treaty Act, as directed by Congress.[36] The list includes 124 species. Except for the mute swan, none has ever enjoyed protection under the MBTA, and most are only casual visitors to the United States. Seventeen of the 124 species have established, self-sustaining populations, including starlings and house sparrows.

Along with the issue of which birds are covered is the question about the range of human activities that fall within the statute's wordy prohibition. The coverage is broad but does have limits, mostly in the case of activities that harm species indirectly and unintentionally.

The Migratory Bird Treaty Act clearly applies to commercial activities, including barter involving bird parts. A typical case, from a federal court in

Tennessee in 2006, involved a performer of Indian dances who was found guilty of possessing and trading feathers from eagles and other migratory birds.[37] He likely could have obtained, but did not, a permit under the MBTA to possess and use the feathers. (The defendant was not a member of a federally recognized Native American tribe and thus, as noted below, could not have gained a permit to possess and use the eagle feathers under the Bald and Golden Eagle Protection Act.) The defendant's conviction was upheld without any involvement on his part in taking the feathers from the wild; his possession and trading were enough.

A starker case was resolved by the federal appellate court in Louisiana in 2002, in *United States v. Morgan,* a decision involving a violation of hunting regulations that limited the number of birds a hunter can possess at any one time.[38] The defendant shot and killed two ducks. When his "poorly trained" dog retrieved six ducks shot by hunting companions as well as his two ducks, the defendant kept the ducks to give to the companions. The court upheld his conviction because the statute banned possession without regard for the reason, at least when the defendant knew he had the ducks.

Another group of cases has considered whether a defendant can violate the Migratory Bird Treaty Act by discharging pollution into water bodies and poisoning birds that visit the water bodies. The issue has produced conflicting interpretations. In a highly visible ruling from 1978, a federal appellate court in New York concluded that a pesticide manufacturer violated the statute by discharging residues from various manufacturing tanks and facilities into a ten-acre settlement pond.[39] Migratory birds visiting the pond died from ingesting the pesticides. The court upheld the conviction of the corporation, citing evidence that it knew the toxicity of the pesticide even if it did not have specific knowledge of the dangers posed to birds. In reaching this conclusion, though, the court rejected the government's claim that the MBTA covered all killings of birds without limitation. It would "offend reason and common sense," the court observed, to include "every killing within the statute, such as deaths caused by automobiles, airplanes, plate glass modern office buildings or picture windows in residential dwellings into which birds fly."

Other courts struggling with indirect killings of birds have shown similar worries. Several have narrowed the statute by ruling that a violation occurs only when a defendant could have avoided the harm through the exercise of due care. Thus in 1978, an aerial applicator of pesticides to farm fields in California was found guilty by a federal district court when the pesticide killed birds, given his negligence in applying the pesticide.[40] A 1989 Idaho case involving similar facts led to acquittal when there was no clear showing of lack of due care.[41] In a 2006 case from Ohio, a steel company was accused of killing birds when chemically laced residues were put into a lagoon visited by birds.[42] The court avoided deciding whether the MBTA applied in the situation because the government had failed to prove beyond a reasonable doubt that the birds died from the contaminants.

Several cases from the 1990s focused on a similar issue: does the Migratory Bird Treaty Act apply to cases in which bird death is due to habitat modification by a landowner? Most of the cases have involved challenges to land management practices by the U.S. Forest Service and other federal agencies. Several cases were resolved in favor of the federal agency on the ground that the MBTA did not apply to employees of the federal government (even though the Fish and Wildlife Service had so applied it for many decades). One court that reached that conclusion, however, also noted that timber harvesting that killed birds could violate the act. The most recent ruling is from the federal appellate court in California in 2004, in a case challenging the cutting of trees with nesting birds by the National Park Service.[43] The court reaffirmed a 1991 ruling from the same circuit that concluded that timber harvesting, and presumably other land uses, did not violate the MBTA—even though the same conduct might violate the Endangered Species Act.[44] These cases were all civil suits brought by plaintiffs seeking to halt unwanted land use practices. None involved a prosecution under the act by the federal government.

Regulations under the Migratory Bird Treaty Act include detailed provisions on hunting migratory birds. They specify seasons; limit the weapons that can be used (mostly to shotguns); ban the use of live decoys, recorded duck calls, and shooting from submerged sinkboxes (all common practices at one time); and prohibit shooting from moving boats or vehicles, among other restrictions. They also specify what species can be taken in different locations. These rules, however, are subject to more-restrictive game laws enacted by the states. Although states cannot reduce federal limitations, they can make them more restrictive. When more protective of birds, states laws govern.

Of all the regulations on taking birds, none has generated more controversy than the ban on hunting over bait. This longstanding regulation promotes sporting ethics and curtails feeding activities that can seriously distort bird behavior. The topic was considered earlier (in chapter 7) in the context of state laws on the subject.

For decades, the Migratory Bird Treaty Act imposed strict liability on hunters for shooting over bait. This meant a hunter could be guilty even if she was unaware that bait was present. The strict liability no doubt proved unfair to hunters. On the other hand, the burden on game wardens and prosecutors would have been excessive if they had to prove beyond a reasonable doubt the hunter's actual knowledge of the bait. A hunter could simply deny knowledge, leaving the prosecutor with a nearly impossible burden of assembling evidence to refute the claim.

Many federal courts were willing to enforce the regulation as written, refusing defendants a chance to deny knowledge. These courts effectively gave prosecutors discretion to refrain from bringing prosecutions in cases in which they believed a hunter in fact had no knowledge of the bait. Other federal courts, however, were more concerned about unfairness. These courts, in effect, amended the regulation to require prosecutors to prove that the hunter either knew about the bait or, with the exercise of reasonable care, should have known about it. The result was what might be termed a "reasonable hunter" standard—would a reasonable hunter, attentive to circumstances, have realized that bait was present?

Congress agreed with this reasoning and amended the Migratory Bird Treaty Act to add the requirement to the statute.[45] Overall, the Fish and Wildlife Service retained very broad discretion to control hunting. But on this one issue, Congress decided itself to lay down the governing rule. It banned the taking of any migratory "game bird" by the use of baiting "if the person knows or reasonably should have known that the area is a baited area." The statute also banned a person from placing or directing the placement of the bait so as to assist in a "take or attempt to take any migratory game bird." The limit to game birds presumably was based on the general illegality of taking nongame birds, with or without bait.

This clarification has substantially reduced the number of baiting cases. Issues about baiting, however, still come up. Factual disputes will always arise about whether a hunter is close enough to the bait to be hunting "on or over" it, as will disputes about the knowledge that hunters possessed or should have possessed. Beyond these points, disputes sometimes arise over "bait" that is present in an area by accident or for reasons unrelated to any attempt to lure game birds. In a prominent 2005 ruling from a federal appellate court, a hunter tried to escape liability on the ground that seed was present in the area because of an accidental spill by a farmer in the course of a normal farming or soil stabilization practice.[46] The court nonetheless upheld the conviction. The prosecution did not have to prove the intent of the person putting the seed in a location. Only the presence of the bait was relevant.

This 2005 ruling touched upon another bait-related issue that is commonly litigated, the exception for seed or other bait present because of farming operations. Many grain growers use harvesting methods that leave considerable grain on the ground. In addition, farmers leave grain in the field into the late autumn and early winter either to let it dry, increasing its value, or because weather conditions or equipment problems have delayed harvest. The question, then, is whether a farmer is leaving crops in a field to serve as bait or whether the seed is present simply as a consequence of ordinary farming methods.

For a time, the law on this subject looked at the intent of the farmer. Was the intent to bait or to engage in normal farming operations? That test proved difficult to apply because it again required prosecutors to prove facts—a person's intentions—that could be shown only indirectly. Current regulations still allow hunters to avoid liability when seed is present due to normal agricultural operations, but the regulations define the protected operations as farming activities "conducted in accordance with official recommendations" of state agricultural extension agents.[47] This qualification imposes a more objective standard; if conduct deviates from the recommendations of state agents, then it falls outside "normal operations" and is covered by the ban on shooting over bait.

A typical case reached the federal appellate court in South Dakota in 2006.[48] A corn farmer allowed his corn to stand in the field into the winter, harvesting only a few rows at a time. By harvesting in this way, the farmer maximized the corn available for geese and encouraged them to return annually and remain on the farm. In addition, the landowner seeded his standing corn with winter wheat. The court concluded that neither practice was a normal agricultural method. Nearly all corn in the area was harvested by December 1, so corn left standing after that date qualified as bait. Similarly, aerial seeding of wheat was not a recommended method and thus qualified as bait. As a result, the use of the corn to attract geese was not impermissible before December 1; the landowner could use corn to lure geese until that date.

A further fact in this South Dakota case highlights another wrinkle in the hunting-over-bait rule. A neighbor of the landowner-defendant also left corn standing in the field after December 1. That corn also was bait, and hunting near it was forbidden. The ban on hunting extended not just to the land of the farmer who left the corn standing but to neighboring land as well. Thus a landowner can be banned from hunting on his land due to activities of neighbors so long as the land is within the "zone of influence" of the bait.

Another issue under the Migratory Bird Treaty Act is whether its regulatory limits apply to actions by federal employees. Surprisingly, the issue has produced sharply contradictory decisions. Several federal courts have held that the act does not apply to federal activities. (Many of these rulings involved challenges to federal timber harvesting activities.) Fish and Wildlife Service regulations, however, have long applied to federal agencies, including employees of the service itself. Agency regulations now give limited exemptions, but their legality is not beyond question. None of the four treaties that the MBTA implements includes an exception for actions undertaken by government agents. The statute thus seems to apply to such employees, notwithstanding either judicial decisions or agency regulations.

This issue drew particular attention in 2002 when a federal court ruled that the Migratory Bird Treaty Act applied to live firing exercises by the U.S. Navy that killed migratory birds.[49] Congress quickly reacted by exempting from the MBTA activities by members of the armed forces during military readiness activities. The statute directed the secretary of defense, in consultation with the secretary of the interior, to "identify measures to . . . minimize and mitigate, to the extent practicable" harms to migratory birds.[50] Importantly, the protective measures were to be written by the military rather than the Fish and Wildlife Service, and the statute did not expressly require the military to follow the protective measures or even to make them public. This exemption for military activities does not appear in the four underlying treaties. For the statute to be viewed as lawful, it must be seen as a partial disavowal of the four treaties by the United States.

A few additional issues under the Migratory Bird Treaty Act merit quick mention. For years, the federal government has largely refrained from applying the MBTA to the use of feathers and other bird parts for ornamental and sacred purposes by members of recognized Indian tribes. It has done so in part because it is unclear whether the MBTA restricted tribal hunting rights (it did not, seems to be the dominant interpretation). Similarly, longstanding disputes have surrounding the taking of migratory birds for subsistence purposes by Alaskan residents.

The Migratory Bird Treaty Act also implicates the private property rights of landowners in ways that merit brief comment. As noted, the hunting-over-bait rules can halt hunting on private land and sometimes on adjacent land. In addition, the Fish and Wildlife Service can ban migratory bird hunting in particular areas. It does so in the case of lands surrounding wildlife refuges, which effectively limits hunting on these lands. This power, of course, is the same as that long possessed and exercised by the states without serious legal challenge. There is no legal reason, based on the Constitution or otherwise, that prevents a federal agency from exercising a regulatory power that states also routinely exercise. The issue has more to do with federal-state cooperation than with private property rights.

Property rights also come up when a landowner kills a bird that is preying on livestock or otherwise causing damage. Regulations provide methods for landowners to obtain permits to deal with birds that cause damage. Landowners who fail to obtain permits, however, have little chance to ward off liability based on a defense-of-property claim. The issue was thoroughly aired in a Massachusetts federal court ruling in 2003 involving a landowner who killed a marauding red-tailed hawk.[51] The court agreed with longstanding precedent that landowners cannot raise defense-of-property to avoid liability under the Migratory Bird Treaty Act. No case, however, has involved the kinds of widespread destruction by wildlife that have led states (as discussed in chapter 4) to let landowners claim defense of property to avoid liability for game law violations in limited circumstances.

Violations of the MBTA require no showing of any particular knowledge or intent by the defendant in the case of misdemeanor penalties—that is, fines of up to $15,000 and imprisonment of up to six months. Felony violations can lead to greater penalties, including imprisonment of up to two years. The felony provisions, however, do not cover all violations of the act. They are limited to purchases and sales (or barters) of birds or bird parts and to takes that occur with an intent to sell or barter. (Only misdemeanor penalties are available for other violations, including takes for personal use, poisonings, killings to protect property, and the like.) In addition, felony convictions require the prosecutor to prove that the defendant engaged in the wrongful conduct "knowingly."

The term "knowingly" was added by Congress in 1986 after several courts expressed resistance to imposing felony penalties without regard for the defendant's knowledge of intent.[52] "Knowingly" clearly requires that the defendant know what he or she is doing—that the conduct be intentional rather than accidental. On the other hand, the defendant need not either intend or desire the harm. "Knowing," that is, is not the same as "willful" or "intentional"; it requires no showing of intent to cause the harm involved. Under the misdemeanor provisions, a defendant who kills a bird cannot escape liability by claiming ignorance that the bird was protected.[53] The same rule is likely true with respect to felony cases. On the other hand, it might be relevant in felony cases whether the defendant knew the species of bird being harmed, at least if defendant claims he thought it was a nonmigratory bird. The Endangered Species Act, which also prohibits takes of species, contains a similar requirement that a wrongful act be done "knowingly" to trigger penalties.[54] Judicial rulings under that statute have disagreed as to whether a mistake regarding the species taken is a valid defense. In a 2007 ruling dealing with the Bald and Golden Eagle Protection Act, *United States v. Zak*, a federal court in Massachusetts ruled that a defendant could not escape liability by claiming he did not know the killed bird was an eagle; that statute similarly penalized any conduct undertaken knowingly.[55]

In the case of the Migratory Bird Treaty Act, the issue of mistaken species might rarely arise. A felony violation requires either an actual or attempted purchase or sale or a take made with intent to sell. A case in which the prosecution can prove this element is unlikely to pose doubts as to whether the defendant actually knew the species involved. In addition, many factual errors are likely to reflect confusion among migratory species. A defendant certainly cannot escape liability by claiming he thought a bird belonged to one migratory species rather than another.

THE BALD AND GOLDEN EAGLE PROTECTION ACT

A third federal wildlife statute, which also directly protects wild species, is the Bald and Golden Eagle Protection Act, enacted in 1940 to protect bald eagles and expanded in 1962 to cover golden eagles. The statute reflects the special cultural status of bald eagles and a national desire to protect them. The basic structure of this statute resembles the Migratory Bird Treaty Act. The Bald and Golden Eagle Protection Act prohibits a wide array of actions that harm eagles and then gives the secretary of the interior power to issue permits. The two statutes differ, however, in many respects.

In terms of actions prohibited, the Eagle Act covers more indirect harms than the Migratory Bird Treaty Act. The key wrongful action is a "take" of an eagle or eagle part. The definition covers not just the expected actions of pursuing, shooting, wounding, killing, and capturing, but also molesting, disturbing, and poisoning. Violations are subject to civil penalties—a fine of up to $5,000 per violation, with each affected bird a separate offense. Because no showing of knowledge or intent is needed, the statute literally prohibits a variety of well-meaning actions that do not actually harm birds. Eagle feathers, of course, have long been collected and incorporated into art, jewelry, and clothing. All sales or exchanges of eagle feathers—indeed their mere possession—violate the act unless a permit is obtained. The civil penalty provision envisions the need for flexibility. It authorizes leniency by the secretary when the facts warrant it. The gravity of the offense and good faith of the violator are relevant, and the secretary can remit or mitigate a penalty as appropriate.

In contrast with the Migratory Bird Treaty Act, all criminal penalties under the Eagle Act require a showing that the defendant acted "knowingly, or with wanton disregard for the consequences."[56] The act originally applied only to conduct that was willful, a standard that made prosecutions difficult and excluded too many actions that killed eagles. When Congress revised the statute in 1972, changing the requirement of willfulness to "knowingly," it also provided that a violation occurred if the defendant acted "with wanton disregard." The effect was to cover cases involving poisoning and other foreseeable takes of eagles that could have been avoided. Few cases have fleshed out the meaning of "knowingly." In a 1975 decision, a federal court in Montana held that the defendant could not escape liability by claiming ignorance of the law.[57] The defendant did not have to "know" about the ban on selling eagle feathers. As an aside, the court stated that "a conviction would not be had were a person to sell golden eagle feathers thinking them to be turkey feathers." In *United States v. Zak,* the 2007 case mentioned at the end of the Migratory Bird Treaty Act section, however, the federal court held that the prosecution need not prove beyond a reasonable doubt that the defendant knew the bird being harmed was an eagle.[58] More broadly, the court asserted that the defendant was guilty "regardless of whether he knew the juvenile bird was an eagle or, as he said, 'a big brown hawk.'" It is possible that courts may in time take a middle position on this issue—not requiring proof beyond a reasonable doubt of the defendant's knowledge, while allowing a defendant to introduce convincing evidence of a reasonable mistake of fact.

The inclusion of the word "poison" in the definition of "take," along with "molest" and "disturb," has encouraged courts to apply the act broadly to actions that harm or kill birds indirectly. A leading federal district court ruling from 1999 upheld the application of the act to an electric utility whose long-distance power lines killed a variety of eagles and other large birds of prey.[59] The court rejected the utility's claim that the Eagle Act was violated only by actions normally associated with hunting or poaching. Citing the expansive definition of "take," the court concluded that the utility was guilty of acting with "wanton disregard" of the consequences because it could have reduced the problem by installing inexpensive equipment but chose not to do so. The court's conclusion was strengthened by a memorandum delivered to Congress in 1972, when the "wanton disregard" language was added. The memo specifically explained that the language was intended to cover electric utilities by requiring them (as well as all other actors) to take reasonable precautions to avoid killing eagles.

The Eagle Act also differs from the Migratory Bird Treaty Act in that the secretary of the interior has almost unlimited flexibility in issuing permits to engage in otherwise unlawful conduct. Indeed, so vast is the flexibility that the act's force is dependent upon political will. Permits can be granted for scientific or exhibition purposes and for the religious purposes of Indian tribes. They can also be obtained when necessary to protect wildlife and "agricultural or other interests in any particular locality." As if these provisions were not broad enough, the statute goes on to authorize the secretary to allow takings of golden (but not bald) eagles "seasonally" to protect "domesticated flocks and herds" when requested to do so by the governor of any state, and to allow taking of golden eagle nests "which interfere with resource development or recovery operations." The broad coverage of the act is thus subject to the secretary's ability to diminish protections significantly. Furthermore, in defining the term "disturb," the Fish and Wildlife Service included an incidental take permit that allows the take of both bald and golden eagles "associated with otherwise lawful activity." These weaknesses are not a new feature. For decades after the original act was passed in 1940, bald eagle populations declined rapidly—a trend that was reversed only after the species was listed under the Endangered Species Act.

A much-litigated issue is the application of the act to Indians and to the use of eagle features in religious ceremonies. The act contains permit provisions authorizing not just possession, but takings of eagles when reasonably linked to religious activities. In a wide-ranging memorandum in 1994, President Clinton directed federal agencies and departments to take steps to assist tribes in acquiring eagle parts and expediting permit processes. Permits, however, are only open to members of registered tribes. Several legal cases have involved claims by sincere believers in Indian religions that they also should have the access to eagle parts, even though not tribal members. The results have been complex and somewhat contradictory.

CHAPTER

10

Wildlife on Federal Lands

Wildlife on federal lands has long drawn considerable attention, and with good reason. The landholdings of the federal government are massive—some 30 percent of the nation's surface—including vast expanses of wildlife habitat. What is the legal status of the wildlife on these lands? Who has the legal authority to manage it? And what are the rules currently in place? These are the questions taken up in this chapter in the context of the four largest categories of federal lands: the wildlife refuges run by the U.S. Fish and Wildlife Service; the variously named holdings of the National Park Service; the national forests; and the diverse lands administered by the Bureau of Land Management. Together these four categories make up over 96 percent of all federal landholdings. The wildlife laws applicable to these lands differ among the four categories and thus require separate consideration. Before exploring them, however, it is useful to understand the basic legal framework that applies on nearly all federal lands.

In 1976, in *Kleppe v. New Mexico,* discussed in chapter 6, the U.S. Supreme Court ruled that the federal government possessed almost unlimited sovereign power to manage the lands that it owned, in addition to its powers as landowner.[1] If it wanted, the federal government could enact its own laws governing all aspects of hunting, fishing, and wildlife management, leaving no role for state laws and state agents. The federal government has not done that, however. State law continues to play a critical role, just as it does in allocating water on federal lands (but not, in contrast, in allocating federal timber, minerals, and forage resources). Nearly all hunting and fishing taking place on federal lands must comply with state licensing requirements, game seasons, bag limits, and the like. States, however, cannot authorize hunting and fishing in places or at times when federal law prohibits it. More generally, state wildlife law has no application when it conflicts with federal statutes and regulations. The federal-state arrangement, in short, is legally complex, and it makes up an initial, major issue of wildlife law on federal lands. What wildlife-related rules has Congress put in place, and how much have they displaced state fish and game laws?

The federal government manages its land by creating land management systems, such as the national forest system and the national wildlife refuge system. Over time, these management systems have come to share a number of characteristics. In most instances, Congress has created an agency to manage the system. The U.S. Forest Service was established to manage lands in the national forest system. Similar roles are played by the Fish and Wildlife Service, the National Park Service, and the Bureau of Land Management (BLM). Congress then specifies some method for designating land to include within a system. Sometimes—as is the case with national forests and wilderness areas—Congress does the designating itself. In other situations, Congress delegates that power. Lands can be placed in the refuge system, for example, by presidential proclamation or federal agency actions. Finally, Congress establishes a set of more or less specific standards to guide the agencies as they manage their lands.

Each of the four major land management agencies is required by Congress to prepare management plans for the systems they control. Sometimes federal statutes give agencies broad discretion in formulating their plans. In other settings, Congress is more specific, requiring plans to evaluate particular topics or forbidding the agency from authorizing particular activities. Within these statutory guidelines, federal agencies then draft their management plans. They present the plans for public comment; prepare environmental impact statements as needed; and ultimately—often after long, frustrating processes—release the plans in final form.

These agency planning processes form a second important component of wildlife law on federal lands. Plainly, the ways lands are managed can greatly affect resident and migratory wildlife—their types, populations, and distributions. Land management, in turn, depends on the goals an agency has for its lands and, in the case of an agency that has conflicting goals, how it sets priorities. This challenge of melding diverse goals is particularly acute for the Forest Service and the BLM, which are required to manage their lands for multiple goals, including wildlife conservation. On these multiple-use lands, wildlife needs somehow must fit together with grazing, timber harvesting, mining, and unrelated recreational activities. Predictably, the legal disputes surrounding such lands are endless, as competing interests squabble. Because courts are typically unwilling to second-guess the substance of agency decisions, disputes that end up in court often center not on actual land management rules, but on issues of power and process—on who makes the rules and how they are made.

These, then, are the two central issues in the study of wildlife on federal lands: how power is allocated between the state and federal governments; and the role of wildlife in agency land management plans. Once that framework is known, the other important pieces usually fall into place easily. On most federal lands, the detailed rules on wildlife harvesting are quite similar to the rules that apply on adjacent private lands. In some settings, federal law does differ, usually by restricting the times and places of harvesting. Rarely, however, do federal agencies promulgate the kind of detailed rules that make state hunting and fishing regulations so bulky.

Two final, interrelated topics also merit quick notice: whether the federal agencies that manage wildlife on their lands can also control wildlife on adjacent, nonfederal lands, to help achieve their land management goals; and whether agencies are obligated to manage their own lands in ways that promote balanced wildlife populations over larger spatial scales.

The survey in this chapter does not examine the Endangered Species Act, which is covered in the next two chapters. Nor does it consider the various ways that conservation interests have used the National Environmental Policy Act, with its well-known environmental impact statement process, to pressure agencies to pay more attention to wildlife and habitat protection.[2] Finally, it does not directly take up the Wilderness Act, which imposes severe limits on the managerial discretion of federal agents in the cases of lands that Congress has designed as wilderness areas.[3] The basic rule in wilderness areas is that extractive land uses are prohibited, along with roads and structures. Regardless of the agency in control, these lands are managed to minimize human interference except for interferences intended to mimic natural disturbances that humans living elsewhere have disrupted.

WILDLIFE REFUGES

The current system of federal wildlife refuges, now over ninety-six million acres, expanded in numerous, unplanned steps over the course of the twentieth century. The first refuge was created by an executive withdrawal by President Theodore Roosevelt in 1903. Some refuge lands were purchased expressly for wildlife protection; some land parcels were donated; and many were shifted into refuge status from a different category of federal landholding, typically from the catchall category of "public domain." Often Congress was the source of the decision to designate a refuge. Just as often, a refuge arose through executive action. For decades, these diverse, fragmented parcels (more than five hundred in all) carried a variety of names and were managed for a wide variety of overlapping goals. They were termed "wildlife ranges," "game ranges," "wildlife management areas," "waterfowl production areas," and "wildlife refuges." In terms of goals, most parcels were acquired or reserved to conserve one or more particular wild species; others were set aside for wildlife more generally. Some were given specific management goals, either by Congress or the president; others were protected with vague or general guidance on management and on how wildlife protection should fit together with other land uses. A great many refuges, particularly along the Mississippi Flyway, encompassed riparian wetlands set aside to protect waterfowl for hunters.

The story since Teddy Roosevelt's day has thus been basically as follows: The refuge system grew haphazardly, parcel by parcel, until Congress around midcentury decided to bring some order to it. In the National Wildlife Refuge Administration Act of 1966, Congress sought coherence by combining hundreds of independent land parcels, subject to varying agency control, into a single system under the control of the Fish and Wildlife Service, in the Department of the Interior.[4] The task of bringing coherence, however, proved quite difficult. One obstacle was the fact that many individual refuges already had legally imposed goals, and they were by no means consistent. In many cases, designation as a refuge required overcoming stiff resistance from groups that wanted to use the lands for grazing, timber harvesting, mining, and other intensive activities. Deals were made; understandings were reached. Politics often produced legal mandates that balanced wildlife protection against conflicting land uses.

Meanwhile, the idea of a game or wildlife refuge itself was changing. In the early twentieth century, wildlife advocates such as William Temple

Hornaday saw game refuges as land parcels that were set aside as no-hunting zones. They were places where game animals could escape hooks, bullets, and traps. The original vision of a federal wildlife refuge—or, at least, refuges acquired to protect migratory birds (as most of them were)—was a place "suitable for use as an inviolate sanctuary."[5] As protected game populations rose, individual animals could wander out of these safe havens and be available for capture on surrounding lands and waters. In the 1930s, conservationist Aldo Leopold, author of the first textbook on game management, proposed a scheme of game refuges widely distributed across the United States to provide breeding stock for hunters everywhere. A hunter himself, Leopold shared Hornaday's view of refuges as no-hunting zones, even as he also saw a growing need for government to provide public hunting grounds.

As the twentieth century unfolded, hunters pressed the federal government to open parts of refuges to hunting; they should be hunting grounds, it was argued, as well as places of refuge. At the same time, conservationists called for refuges to serve as habitat for nongame species, including threatened and endangered ones. Sometimes this mix of game and nongame species could be accomplished without interfering with game management goals. But conflicts did arise. By the 1980s, many conservationists were pushing to have refuges managed strictly as natural areas, with as little human alteration as possible. They opposed manipulating lands to increase populations of particular species (usually ducks) at the expense of other, less-glamorous species. On many refuges, managers had intentionally increased waterfowl populations by means of artificial ponds and altered vegetation. Typically this policy was intended to compensate for habitat losses the species had suffered elsewhere. But many conservationists were becoming uneasy with this kind of manipulation, however well-meaning. They preferred to see land managed to promote the full range of species that would be present in a place had humans not intervened.

Conflicts over goals delayed Congress's efforts to turn refuges into an orderly system. But progress slowly occurred. A key step came in 1962 with the enactment of the Refuge Recreation Act.[6] The act introduced what became the central ordering principle for refuge management: the idea that a refuge has one or more primary purposes and that activities unrelated to the primary purposes are permissible only if "compatible with" them. The 1962 act authorized a wide range of recreational activities, subject to this new compatibility standard. By then, hunting had become one of the most popular wildlife-related recreations. Earlier, in 1949, Congress had authorized refuge managers to open up 25 percent of each refuge to hunting, chiefly as a concession to duck hunters, who faced an increase from one dollar to two dollars in the price of duck stamps (the primary source of funding for refuge acquisition).[7] Coincident with a further price rise in 1958, Congress increased the 25 percent figure to 40 percent (where it remains today, subject, as noted below, to further expansion when needed to curb excessive populations).[8]

A more significant step toward systemwide coherence occurred in 1966, when Congress enacted the Refuge Administration Act, which organized refuges into the national wildlife refuge system and supplied additional nationwide guidance.[9] The statute directed the secretary of the interior to manage refuges "to conserve and protect migratory birds . . . and other species of wildlife found thereon." It also authorized the secretary to "manage timber, range, and agricultural crops" and to permit uses of refuges for "any purpose" so long as it was compatible with the individual purposes of the individual refuge and with the new systemwide goal of wildlife conservation. The 1966 statute contained a dominant-use management standard by requiring compatibility determinations before permitting nonwildlife uses of refuge lands. Embedded in the legal scheme, however, was a tension. How were managers to handle individual refuges when their specific charters authorized or even directed land uses that conflicted with nationwide wildlife goals? What did compatibility mean when the goals themselves were not compatible?

Although the 1966 statute increased the coherence of the system, dissatisfaction nonetheless surrounded its broadly worded management goals and its lack of planning requirements. After years of further hearings and countless legislative proposals, Congress in 1997 finally achieved its restructuring aim. It gave to the refuge system a detailed charter, similar to the ones it had given the Forest Service and Bureau of Land Management in the 1970s. The 1997 National Wildlife Refuge System Improvement Act clarified the system's overall mission, provided guidance for making compatibility determinations, and explained how the goals of individual refuges fit together with the goals of the system as a whole.[10] In addition, the act set clear priorities among compatible uses when, as often, managers had to choose among them.

The current mission of the refuge system is "to administer a national network of lands and waters for the conservation, management, and where appropriate, restoration of the fish, wildlife, and plant resources and their habitats within the United States for the benefit of present and future generations of Americans."[11] Thus wildlife habitats as such, including plants, are within the system's conservation goal, not just wild animals. The statute further provides that the secretary of the interior is to manage the system to maintain its "biological integrity, diversity, and environmental health." This language instructs refuge officials to manage lands to conserve the full range of wild species and plant communities that were present in a refuge before humans changed the landscape substantially. It also calls for the conservation of basic ecological processes with little human alteration, including the natural biological processes that shape genomes, organisms, and communities. (The Fish and Wildlife Service added detail to this pro-preservation standard in guidelines released in 2001.)[12] This overall system guidance, however, is subject to a distinctly important limitation: refuge managers are also directed to achieve the diverse goals of the individual units set by statute or executive order. In the event these individual-unit goals conflict with overall system goals, priority goes to the goals of the individual refuge.

The hundreds of units in the system, as noted, were typically added one by one—using money from various sources, by a variety of executive and legislative actions, and under the authority of various statutes. Many units today possess individual goals phrased entirely in terms of wildlife conservation, sometimes referring to all wildlife; quite often to birds, native birds, or migratory birds; sometimes to threatened and endangered species; and sometimes to one or more identified bird species. Occasionally, other specific animals are given primacy (for instance, the Key deer in Florida). Other refuges remain guided by express instructions to allow the continuation of—or even to require the promotion of—nonwildlife uses of the land (grazing or mineral development, for instance) that existed when the refuges were created.

A look at a few individual refuges can illustrate this diversity of goals and thus the challenges faced by individual refuge managers.[13] The Clear Lake National Wildlife Refuge in California was set aside in 1911 to serve "as a preserve and breeding ground for native birds" and is "dedicated to wildlife conservation . . . for the major purpose of waterfowl management." Managers of Clear Lake, though, are also required to give full consideration to "optimum agricultural use that is consistent therewith." The Kofa National Wildlife Refuge in Arizona was "set apart for the conservation and development of natural wildlife resources, and for the protection of public grazing lands and natural forage resources" with the express order that "all of the forage resources in excess of that required to maintain a balanced wildlife population . . . be available for domestic livestock." The Crab Orchard National Wildlife Refuge in Illinois was set aside for the conservation of wildlife and for "the development of the agricultural, recreational, industrial, and related purposes."

Bringing coherence to such conflicting goals poses knotty challenges. To cite one example, confusing language in the executive order creating the Charles M. Russell National Wildlife Range in Montana led to protracted litigation. In 1983, a federal court finally decided that the executive order chartering the refuge required that forage resources be used to sustain 400,000 sharptail grouse, 1,500 antelope, and sufficient other species as needed to maintain a balanced wildlife population.[14] All additional forage on the refuge would be shared equally between livestock and wildlife.

Once goals for an individual unit are sorted out, managers are instructed by the Refuge Improvement Act to then decide what human uses of a refuge are appropriate.[15] Human uses, other than those described in the refuge's mission, are divided into two categories. Priority is given to the category termed "wildlife-dependent recreation," which includes hunting, fishing, wildlife observation, and ecological study. All other uses (grazing, timber harvesting, other recreation) are placed in a second category with lower priority. Wildlife-dependent activities, according to the statute, are a "legitimate and appropriate general public use" of refuges. These activities are to receive "priority consideration in refuge planning and management" and "enhanced consideration over other general public uses." Refuge managers still must determine when and whether wildlife-dependent recreational activities are compatible with refuge goals and with public safety. Further, hunting remains limited to no more than 40 percent of a refuge unless the refuge manager determines that hunting would be "beneficial to the species." However, "no other determinations or findings are required" for a refuge official to allow wildlife-related activities to take place.

In the 1997 act, Congress also directed refuge officials to prepare "a comprehensive conservation plan" (CCP) for each refuge or related complex of refuges.[16] Once approved, a CCP governs activities within a refuge until it is revised. A plan must state the applicable goals and describe "the distribution, migration patterns, and abundance of fish, wildlife, and plant populations and related habitats." It must identify the "significant problems that may adversely affect the populations and habitats of fish, wildlife, and plants within the planning unit and the actions necessary to correct or mitigate such problems." Finally, the CCP needs to prescribe opportunities for compatible wildlife-dependent recreational uses.

Land management decisions by refuge managers have generated comparatively little litigation, or at least have yielded few reported decisions in which courts probe the discretion that refuge managers wield. A 2006 ruling upheld the discretion of the Great Dismal Swamp National Wildlife Refuge in Virginia to authorize a hunting season for black bears.[17] The management plan limited the hunt to a remote 20 percent of the refuge that was otherwise closed to users. Uncontradicted research by the refuge concluded that the hunt would have only nominal effects on the bear population. Similarly, a federal appellate court in 2004 upheld the decision of a refuge manager in Nebraska to impose a moratorium on further permits for commercial canoeing outfitters because rising canoe use of the refuge seemed to disrupt bird behaviors, including reproduction.[18] The court stated that it could overturn an administrative decision only if evidence showed the decision was arbitrary and capricious. In another litigation involving a compatibility determination, a federal court in 1978 struck down a decision to allow recreational boating and water skiing in the Ruby Lake National Wildlife Refuge in Nevada.[19] The court stated that agency's decision—that the recreation was compatible—was arbitrary and capricious in light of the substantial evidence that boating and skiing were reducing waterfowl populations.

A more probing inquiry into a refuge manager's discretion was conducted by a federal court in 2007.[20] The case involved a claim by a county cattlemen's association that managers of a forty-thousand-acre refuge in eastern Washington, originally set aside for wildlife, erred by limiting livestock grazing. The grazing was largely ended (after more than sixty years) because refuge managers viewed it as incompatible with wildlife goals. The plaintiff-grazers claimed that a curtailment of grazing could occur only after the agency conducted site-specific, scientifically controlled studies to determine grazing's exact effects. The court disagreed. Refuge officials were expressly authorized by statute and regulations to make compatibility decisions based on their "sound professional judgment," which meant a judgment "consistent with principles of sound fish and wildlife management and . . . available science and resources." No more detailed study was legally required.

To the extent that a refuge allows hunting and fishing, these activities are governed by state fish and game laws as well as federal regulations. As noted, the federal government possesses full power to manage its land and resources and can choose to displace state laws. Accordingly, it is up to Congress to decide the roles that state law will play within refuges. Decisions by the Fish and Wildlife Service can also raise this issue since a state game law remains valid until it actually conflicts with a management plan or an authorized action of a refuge manager.

In the case of wildlife on federal lands, various statutes express Congress's strong desire that the Fish and Wildlife Service work cooperatively with states when managing wildlife on refuges. The main charter for the national wildlife refuge system—the 1966 Refuge Administration Act as amended in 1997 by the Refuge Improvement Act—states expressly that the act should not be construed to affect the "authority, jurisdiction, or responsibility" of the states "to manage, control, or regulate fish and resident wildlife under State law or regulations in any area within the System."[21] Federal regulations "permitting hunting or fishing of fish and reside wildlife" must be, "to the extent practicable, consistent with State fish and wildlife laws, regulations, and management plans." Another provision directs refuge planners to make their conservation plans, "to the extent practicable, consistent with the fish and wildlife conservation plans of the State in which the refuge is located." Further sections call for "effective coordination, interaction, and cooperation," while reaffirming the basic role of the States in fish and resident wildlife management.[22]

Despite their sentiment and clarity, these various provisions do not ultimately limit the ability of refuge managers to carry out their missions. Nor do they empower states to resist a lawful federal action since a state law cannot interfere with the accomplishment of a federal objective. States can insist on being consulted, but it is up to federal managers to decide when and whether it is "practicable" for state law to remain in force. In the case of conflict with a lawful federal action or policy, state law must give way.

This issue of state power was considered in two federal appellate court rulings in 2002. In one ruling, the court concluded that Wyoming officials had no power to enter the National Elk Range to vaccinate elk with brucellosis vaccine.[23] Although state authority, the court concluded, did extend to the refuge generally, federal regulations preempted it "to the extent the two actually conflict, or where state management and regulation stand as an obstacle to the accomplishment of the full purposes and objectives of the Federal Government." In the case of the elk refuge, federal officials viewed elk vaccination by the state as inconsistent with federal elk management efforts.

Later the same year, another federal appellate court embraced the same understanding of federal-state powers.[24] The dispute surrounded Proposition 4, approved by California voters in 1998. The proposition banned the use of certain traps and poisons to capture or kill wildlife in the state, including wildlife on federal lands. Refuge managers used control measures banned by Proposition 4 to protect birds from predation. Although the court noted that the voter-approved state laws generally applied on federal refuges, it nonetheless held that the ban did not restrict federal agents in the exercise of their management activities.

Since 1948, the primary statute protecting refuges has been the Refuge Trespass Act.[25] In addition to hunting, trapping, capturing, and killing "any bird, fish, or wild animal of any kind whatever," the statute also bans willfully disturbing them or taking or destroying eggs or nests of any bird or fish. Activities covered by this ban are permissible only "in compliance with rules and regulations promulgated by authority of law"—which is to say, state and federal law in combination. The 1966 Refuge Administration Act as amended in 1997 by the Refuge Improvement Act includes further prohibitions and penalties: no person may "take or possess any fish, bird, mammal, or other wild vertebrate or invertebrate animals or part or nest or egg thereof," unless otherwise authorized by law. The terms "fish" and "wildlife" include "any wild member of the animal kingdom whether alive or dead, and regardless of whether the member was bred, hatched, or born in captivity." All violations of the statute or implementing regulations are subject to fines and imprisonment up to six months. Stiffer penalties are imposed on violations that are undertaken "knowingly."[26]

The bans on taking or harming wildlife clearly apply only on the refuges themselves. Although the federal government possesses power to protect federal properties against external threats, including land uses that interfere with federal goals, the statutes governing wildlife refuges give refuge managers no power to protect wildlife outside refuge borders. Indeed, several provisions expressly admonish refuge managers to leave private lands alone. The act states in one section that nothing in it "shall be construed to authorize the Secretary to control or regulate hunting or fishing of resident fish and wildlife on lands not within the system."[27] The words are repeated almost verbatim in another provision of the same statute, this time referring to "fish and resident wildlife on lands or waters that are not within the System."[28] Plainly, Congress was pressured to ensure that refuge managers did not ban hunting on adjacent lands, as they had done before and might still do under the authority of other federal statutes.

These various statements by Congress, though, should not be taken out of context. Federal law still overrides state law. Federal agents can still carry out their duties as needed, without having to comply with state law. The issue of federal power outside reservation boundaries was probed in a 2006 federal appellate ruling arising out of Wyoming.[29] Wyoming officials arrested employees of the Fish and Wildlife Service for entering private land without permission to carry out a program of collaring wolves for radio tracking. The state charged the employees with trespass and littering. The appellate court dismissed the charge on the ground that the employees were protected by an immunity arising out of the supremacy clause—the clause in the Constitution that makes federal law supreme over state law in the case of conflict. This federal immunity applied so long as the employees had authority for their actions under federal law and so long as they had a well-founded, objectively reasonable basis to believe that their actions were necessary to carry out their duties.

Also relevant to this issue are the many statements by Congress requiring the secretary of the interior, as head of the national refuge system, to look beyond refuge boundaries when planning and managing refuges.[30] One of the 1997 Refuge Improvement Act's most prominent elements was its instructions to the secretary of the interior to take into account the ways that wildlife refuges might promote wildlife populations, and sustain ecological processes, in larger landscapes. Indeed, the mission of the entire system is to maintain wildlife and their habitats "within the United States," not just within refuge boundaries.[31] Similarly, the instruction to maintain "biological integrity, diversity, and environmental health" applies to the system as a whole, not to individual parcels in isolation.[32] Another statutory provision requires the secretary to plan the system's expansion "to contribute to the conservation of the ecosystems of the United States"—another goal that disregards refuge boundaries.[33] Even more explicit are the statements by the Fish and Wildlife Service itself, contained in its guidance implementing the "biological integrity, diversity, and environmental health" mandate. This guidance aims to promote not just "fish, wildlife, and habitat resources found on refuges," but also those on "associated ecosystems."[34] The three key terms themselves are to be described "at various landscape scales from refuge and ecosystem, national, and international."[35] Further, according to the agency guidance, the desirability of promoting these goals at larger spatial scales can sometimes justify actions within a refuge that might "compromise elements of biological integrity, diversity, and environmental health at the refuge scale."[36] Although satisfying these goals within refuges ordinarily will come first, the system's goals also include restoring "lost or severely degraded elements of integrity, stability, [and] ecosystem health" at much-larger landscape scales.[37]

What these and other legal provisions make clear is that Congress and national wildlife refuge system officials expect individual refuge managers to look over the fence. They should do so not simply so that refuges can be ecologically healthy, but because a refuge's health is integrally linked to that of nearby lands. The Fish and Wildlife Service as a whole, in short, is expressly charged with enhancing ecological conditions on a wide variety of non-refuge lands.

In all likelihood, fulfilling these border-transcending duties is subject to the discretion of national wildlife refuge system officials and the heads of individual refuges. These duties do, however, shape the ways refuge officials ought to think about their functions. The overall aim is no longer simply to provide a sanctuary for game so that hunters have something to shoot. It is to enhance the full range of life on refuges and in the landscapes that include them—a distinctly lofty, challenging, and inspiring management goal.

NATIONAL PARKS

The national park system, like the national wildlife refuge system, includes a large number of separated land parcels, located from coast to coast and of widely varied size. Units managed by the National Park Service lack the consistent naming pattern that the refuge system has finally achieved. Thus the Park Service manages not just national parks but also national grasslands and seashores, national parkways and monuments, national historic sites, national recreation areas, and many others. The service divides these units into three basic categories—natural, recreational, and historic. Overall, the service controls nearly eighty million acres.

Since its founding in 1916, the National Park Service has been guided by a brief, two-part management standard.[38] It is charged with promoting and regulating the parks for public enjoyment—that is, recreation—while also keeping them "unimpaired for the enjoyment of future generations." Among all human uses of land, recreation tends to disrupt nature less than most others. But it can still be disruptive, particularly in its intensive forms and when an agency manipulates nature to increase its aesthetics and to ensure that visitors face few dangers from wild animals. Roads and visitor amenities can fragment habitat and interfere with migrations. Predator control efforts can disrupt the composition of biological communities. Indeed, the National Park Service over the decades has faced charges of manipulating nature so as to create sweeping vistas and to elevate populations of the kinds of safe, attractive species that visitors like to see. For decades, roadways were manicured to produce a familiar "national parks" look. Predator control and fire suppression were widespread. Hay was fed to elk and garbage to bears, with bleachers constructed so visitors could watch. For every wildlife scientist who spoke out for ecologically balanced wildlife populations, the Park Service hired literally dozens of landscape architects to make parks attractive for visitors.

Much has changed since the 1960s. The chief goal remains what it has long been, to enhance the pleasure of millions of park visitors and to encourage more people to come. At the same time, the Park Service is now more inclined to provide habitat for all species, to sustain ecological processes, and to restrict visitor access and even safety when required to protect species at risk.

In terms of wildlife law, the federal-state arrangement on Park Service lands is similar to the arrangement on national wildlife refuges. State law applies except to the extent it is displaced by federal statutes, regulations, and managerial actions. Federal statutes governing the Park Service authorize it to work closely with the states, including (by implication) fish and game officials. But the governing statute is much more vague than in the case of the national refuge system. Cooperation with states is authorized, not mandated, and there are no express references to coordinating federal laws with state fish and game laws. The difference arises in part because most units managed by the Park Service are closed to public hunting and trapping, including the expansive, well-known parks in the lower forty-eight states. In addition, wildlife management does not play the same dominant role in national parks that it does in national wildlife refuges. Nonetheless, current Park Service policy expressly incorporates state fish and game laws. When hunting, fishing, and trapping are allowed in park units, they generally must be undertaken in compliance with state laws, so long as the state laws do not conflict with federal statutes and regulations.

Wildlife law on Park Service lands largely centers on two issues: When are particular park units open to wildlife harvesting, and of what type? And what obligations does the Park Service bear to manage its lands to promote wildlife populations? On both issues, the answers are shaped by the broad statutory guidelines set forth by Congress, particularly the two-part mandate to promote recreation while keeping lands unimpaired. Also influencing the answers are the specific statutes and other orders that created particular park units. Like the national wildlife refuge system, the national park system is the result of many individual actions by Congress and various executive orders. Each unit received an overall goal. Many times, the goal of a park unit expressly authorizes or even mandates hunting, fishing, and trapping. By statute, managers of the individual units are required to fulfill these unit goals, along with the goals set for the Park Service as a whole.

After decades of some uncertainty, the Park Service in the 1980s finally embraced a clear stance on the ability of citizens to hunt, fish, and trap on lands managed by the service.[39] Recreational fishing is generally allowed, subject to normal rules and regulations that are promulgated by the service to supplement state fishing rules. Commercial fishing is allowed only in unusual circumstances when expressly authorized by federal statute. (In a prominent 1997 ruling, a federal appellate court concluded that the Park Service could authorize commercial fishing in Alaska's Glacier Bay National Park generally, but not in areas of the park designated by Congress as federal wilderness.[40]) In all settings, the Park Service has discretion to regulate fishing to protect fish populations.

In the view of the Park Service, hunting and trapping are permissible in park units only when specifically authorized by Congress. When creating many individual park units, Congress occasionally directed the Park Service to permit these activities. In other units, Congress authorized the activities but did not mandate them. When Congress has made the decision itself, a unit is open to hunting subject to reasonable agency regulations and compliance with state law. When Congress has merely authorized hunting, then Park Service policy allows hunting if the superintendent of the unit "determines that such activity is consistent with public safety and enjoyment, and sound resources management principles." The validity of this overall legal interpretation was challenged in 1986 by the National Rifle Association.[41] A federal court sustained the Park Service position as a reasonable interpretation of the applicable federal statutes. In so doing, the court distinguished between hunting and trapping. It agreed with the Park Service, as did a later court, that trapping was permissible only when Congress expressly authorized it by name.

The issue of trapping was the subject of a separate legal challenge in 1991 concerning Michigan's Pictured Rocks National Lakeshore and

Sleeping Bear Dunes National Lakeshore.[42] Congress had directed that the units be open to hunting and fishing but had made no mention of trapping. Park Service policy, then as now, allowed trapping only when expressly authorized by Congress. A federal appellate court upheld the policy on the ground that the Park Service was obligated to conserve wildlife except when told to do otherwise by Congress.

In these two judicial rulings, the federal court upheld the Park Service's interpretation of federal statutes to prohibit hunting and trapping except when congressionally authorized. In so doing, the courts gave deference to the Park Service, upholding the service's legal interpretations on the ground that they were reasonable and thus permissible. In neither instance did the court expressly hold that the Park Service was *obligated* to interpret the statutes as it did. In other words, the courts did not rule out the possibility that the Park Service could have interpreted the statutes differently.

The idea of hunting in national park units continues to arouse public resistance and occasional litigation. Cases commonly arise when hunting is authorized by Congress in a particular unit, but not mandated. In such instances, the Park Service has discretion as to whether to allow it. In exercising its discretion, the service must comply with all laws, including its overall obligation to conserve the parks and keep them unimpaired for future generations. Given its duty to avoid impairment, could the Park Service act unlawfully by not allowing hunting in a particular unit, even when Congress has authorized it?

This legal issue arose in a 2003 case in which an animal welfare group challenged a bear hunt proposed in the Delaware Water Gap National Recreation Area.[43] The statute creating the recreation area contained an extensive provision dealing with hunting and fishing. In it, Congress stated that hunting and fishing should be permitted. Congress went further, however, to state that the secretary of the interior could designate no hunting zones "for reasons of public safety, wildlife management, administration, or public use and enjoyment not compatible with hunting." The bear hunt was sponsored and regulated by the state; the Park Service merely reviewed the hunting plan and expressed no objection to having portions of the recreation area included in the hunt area. The federal court ruled that the Park Service's action did not violate its duty to manage its lands for recreation without impairment. The court also ruled that the service had no obligation to promulgate regulations authorizing the bear hunt: it could simply stand aside while the state conducted the hunt.

By and large, parties challenging actions by the Park Service have had little success in claiming that a specific action by the service violates its obligation to manage its lands without impairment. The National Park Service Organic Act of 1916, which sets forth the dual mandates of promoting and regulating lands for both recreation and nonimpairment, creates a strong tension.[44] When it comes to resolving this tension—deciding when recreation can occur, despite some level of impairment—courts have routinely deferred to the Park Service's expertise. In a 1987 case, a federal district court upheld a Park Service decision to keep open a public campground in Wyoming, even though campers were interfering with grizzly bear behavior.[45] In another case, a court in Massachusetts refrained from second-guessing the service when it decided to continue allowing off-road vehicle access to a seashore.[46] In a 1999 ruling, a federal court in California upheld the Park Service's decision to reconstruct a storm-damaged road, despite alleged harm to nesting bats and rainbow trout.[47] And, in a 1998 ruling, a federal district court invalidated a plan by the Park Service to continue allowing four-wheel-drive vehicles to have access to rough jeep tracks and trails in Utah's Canyonlands National Park.[48] Based on the factual record, the court held that continuance of off-road travel was "inherently and fundamentally inimical" to the continued existence of important, rare riparian areas. This trial court ruling, however, was overturned on appeal.[49] In the view of the appellate court, the ideal of no impairment was inherently ambiguous. The Park Service was empowered in its discretion to balance impairment "against the other value of public use of the park." In so ruling, the appellate court concurred in what is the dominant view, that the Park Service has broad, if not unlimited, discretion in how it goes about balancing recreation and nonimpairment under its statutory charter.

THE MULTIPLE-USE LANDS OF THE NATIONAL FOREST SERVICE AND THE BUREAU OF LAND MANAGEMENT

A full 20 percent of the United States—some 455 million acres—is managed by two federal agencies, the U.S. Forest Service, a unit of the Department of Agriculture, and the Bureau of Land Management, a part of the Department of the Interior. Although the two agencies are governed by different statutes—the Forest Service by the National Forest Management Act,[50] and the Bureau of Land Management by the Federal Land Policy and Management Act[51]—the agencies are subject to broadly similar requirements. The land for which they are responsible are managed for multiple, conflicting uses, including wildlife. Both agencies are obligated to prepare land use plans that take wildlife into account. And the lands of both agencies are almost entirely subject to state fish and game laws and state wildlife management. Given these similarities, the two agencies are usefully considered together. The key difference between them is that the Forest Service is expressly obligated in its land use plans to promote the diversity of wild species; the duty of the Bureau of Land Management to do so is more vague. These statutory provisions and the agency regulations issued under them have engendered considerable litigation.

The National Forest Service

The basic charter of the Forest Service dates from 1897. In the Organic Act, Congress provided that forest reserves could be set aside for three, related purposes: to "improve and protect forests"; to secure "favorable conditions of water flows" (that is, to protect watersheds from eroding and degrading waterways); and "to furnish a continuous supply of timber."[52] By the 1920s, the Forest Service was affirmatively managing its lands for wildlife-related recreation, particularly hunting, as it sought to compete with the more recent National Park Service to attract recreational visitors. In 1960, Congress revised the official mission of the Forest Service, putting into law what the Forest Service had essentially been doing on its own. The 1960 Multiple-Use Sustained-Yield Act stated that national forests should be administered "for outdoor recreation, range, timber, watershed, and wildlife and fish purposes."[53] These multiple uses were supplemental to the original purposes of the forests—that is, supplemental to the Forest Service's duty to "improve and protect" the forests.

According to the 1960 statute, multiple use meant managing the various resources on the forests "so that they are utilized in the combination that will best meet the needs of the American people." Going further, this meant the "harmonious and coordinated management of the various resources, each with the other, without impairment of the productivity of the land, with consideration being given to the relative values of the various resources, and not necessarily the combination of uses that will give the greatest dollar return or the greatest unit output." Multiple use also meant that the Forest

Service had to give "due consideration . . . to the relative values of the various resources in particular areas." In a different provision, the 1960 act authorized the Forest Service "to cooperate with interested State and local governmental agencies"[54] and made clear that nothing in the act should be construed "as affecting the jurisdiction or responsibilities of the several States with respect to wildlife and fish in the national forests."[55]

The multiple-use ideal provides the overall vision for national forest management. As a legal mandate, however, it is sufficiently vague to give the Forest Service substantial if not unlimited discretion in making trade-offs among the multiple uses. Very few courts have had occasion to hear legal claims that the Forest Service has unduly favored one resource (usually timber) over others. The prevailing interpretation is that the Forest Service complies with this act so long as it gives "due consideration" to the various resources, even if it then proceeds to favor one or more of the resources at the expense of the others.

In the 1970s, Congress enacted two major statutes that gave the Forest Service a detailed framework to use in making land use plans. The National Forest Management Act, enacted in 1974 and 1976, requires the Forest Service to prepare regular plans, both for the forest system as a whole and for each forest individually.[56] The planning is to be done using "a systematic interdisciplinary approach to achieve integrated consideration of physical, biological, economic, and other sciences." The statutes imposed specific limits in a number of areas, particularly methods of timber harvesting. In the critical section relating to wildlife, the Forest Service is instructed to provide in its forest plans "for diversity of plant and animal communities based on the suitability and capability of the specific land area in order to meet overall multiple-use objectives."[57] This section, commonly termed the "diversity provision," requires the Forest Service to pay special attention to wildlife, beyond what would be necessary under the vague multiple-use mandate. It requires efforts to promote wildlife diversity as part of the service's larger effort to foster multiple uses.

The Forest Service implemented this diversity provision by issuing regulations that went well beyond the statute.[58] Its regulations called for greater protection of wild species, and set specific rules for how forest managers should achieve it. One regulation stated that "fish and wildlife habitat shall be managed to maintain viable populations of existing native and desired non-native vertebrate species." This protection of viable populations required forest planners to designate sufficient habitat "to support, at least, a minimum number of reproductive individuals," with the habitat "well distributed so that those individuals can interact with others in the planning area." A related regulation called for "management prescriptions, where appropriate and to the extent practicable" to "preserve and enhance the diversity of plant and animal communities, including endemic and desirable naturalized plant and animal species, so that it is at least as great as that which would be expected in a natural forest."

To carry out this diversity mandate, the regulations required forest planners to select, for each forest, "management indicator species" that would become the focus of wildlife conservation efforts. Only a few species would be chosen as indicator species because of the impossibility of trying to manage lands for hundreds of species individually. In selecting the management indicators, however, forest planners were obligated to include threatened and endangered species; species with "special habitat needs" that could be harmed by normal forest uses; and species whose population fluctuations could usefully inform forest planners as to the effects of management activities on a range of other species and on water quality.

Conservation interests of various types soon seized upon these Forest Service regulations to challenge both land use plans and individual actions because of their alleged ill effects on various wild species. Numerous cases challenged the particular species that were chosen as management indicator species, as well as the alleged failure of the Forest Service to study the species with sufficient care and continuity. With respect to the species chosen as indicators, a common complaint was that the Forest Service too often chose species that could thrive in landscapes fragmented by timber harvesting, at the expense of species that required unbroken forests, particularly "old growth" forests. Courtroom clashes also occurred as to how many animals were needed to sustain a population's viability, and whether the Forest Service, in determining viability, could take into account animals residing on adjacent lands not controlled by the service.

Another recurring question was whether the Forest Service could adequately protect the diversity of wildlife by ensuring protection of a diversity of vegetative communities—that is, whether it could simply protect habitat diversity and assume that diverse animal species would inhabit them. Because habitat protection often had to be planned at large spatial scales, many of the lawsuits challenging the Forest Service entailed challenges to the plans for entire forests, rather than challenges to particular proposed timber cuts or road-building projects. For forest planners, this litigation proved frustrating because it forced them to make more fine-grained land use decisions at the level of the entire forest, a difficult challenge, given the need to compile massive amounts of information. It also produced plans that were relatively inflexible, given that a plan, once effective, had the force and effect of law and was binding on all subsequent individual decisions made within the forest.

The most prominent decision arising under the diversity provision in the statute and its implementing regulations arose in Wisconsin in 1995.[59] In the case, *Sierra Club v. Marita,* the Sierra Club challenged the plans for several entire forests, claiming that the Forest Service failed to structure its plans based on the principles of the scientific field of conservation biology. The plaintiffs complained that, while the Forest Service designated and protected considerable wildlife habitat within the forests, it failed to give due weight to the sizes of habitat "patches," to their spatial distribution, and to the need for unbroken corridors to connect them, thereby allowing individual animals to travel safely between the patches. Both district and appellate courts disagreed with the plaintiffs' interpretation of the statute and regulations, even while acknowledging the scientific soundness of their claims. As the court interpreted the relevant legal provisions, they did not compel the Forest Service to embrace any particular scientific principles or theories. The service was free to implement its regulations as it saw fit, so long as its decisions were not arbitrary and capricious.

In other cases, environmental plaintiffs often prevailed. Courts pushed the Forest Service to pick different indicator species, to monitor their populations more closely, to do further research on minimum population sizes, and to pay attention to particular habitat needs. In several important cases, courts ruled that the Forest Service could not undertake timber harvesting actions without first performing actual detailed studies of the populations of indicator species; it could not simply assume that populations were adequate based on the presence of suitable habitat.

The Forest Service made major changes in this regulatory regime in 2005.[60] The new regulations replaced the previous diversity requirements and significantly reduced the Forest Service's obligation to protect species diversity on forest lands. In combination with two important Supreme Court cases from about the same time, it also reduced the ability of conservation groups to challenge forest planning at the whole-forest level. The 2005 rules represented, according to the Forest Service, a "paradigm shift in land management planning."[61] Under the old rules, forest plans incorporated many key policy decisions. They laid out the framework for managing particular forests and constrained the choices that forest managers could make on a day-to-day basis. Under the new rules, forest plans were more aspirational and goal oriented. The plans addressed broader issues—desired forest conditions, overall objectives, and so on—but without making any specific commitments or final decisions. The Forest Service's aim in revising its regulations was to halt nearly all litigation that challenged forestwide planning. It wanted to limit citizens to challenging only specific actions taken on the ground by managers of particular forests—individual timber sales, for example.

One effect of this new approach was to make it difficult, if not impossible, for conservationists to do what the Sierra Club was able to do in *Marita:* to challenge forest planning at the whole-forest level. In taking this approach, the Forest Service drew upon two important rulings by the United States Supreme Court—*Ohio Forestry Association v. Sierra Club* (1998)[62] and *Norton v. Southern Utah Wilderness Alliance* (2004),[63] which significantly curtailed the ability of litigants to challenge federal land use plans in the absence of a specific action, taken on the ground, that physically altered nature in ways that harmed the plaintiffs.

The 2005 regulations eliminated the various regulatory sections dealing with biodiversity protection, including the regulations requiring management indicator species and mandating that "fish and wildlife habitat . . . be managed to maintain viable populations of existing native and desired nonnative vertebrate species." The new regulations instead called for forest management based on the notion of sustainability.[64] "Sustainability" is defined to include sustaining the "social and economic systems" in forest areas—meaning, presumably, the full range of existing extractive forest uses. It also includes an ecological element: forest plans will "provide a framework to contribute to sustaining native ecological systems by providing ecological conditions to support diversity of native plant and animal species."[65] The "primary means" of promoting such ecological systems is to promote "ecosystem diversity." Thus forest plans thereafter will "establish a framework to provide the characteristics" of ecosystem diversity. Unless the official responsible for a particular forest decides otherwise, nothing further will be done to carry out the obligation in the National Forest Management Act to provide for "diversity of plant and animal communities." If the responsible official thinks it necessary, a forest plan could include additional provisions "to provide appropriate ecological conditions" for particular categories of species, including threatened and endangered species, "species of concern," and "species of interest."[66] Any such additional provisions, however, must be consistent with overall multiple-use objectives.

These 2005 rules essentially shift planning attention away from the protection of species as such to the conservation of ecological conditions. Under them, the manager of a forest can now focus entirely on conserving vegetative communities and need not identify individual species nor monitor population levels and trends. Moreover, "diversity" seems to mean simply variety, so that a forest can satisfy the diversity requirement even while losing various types of biotic communities. The effect is to give forest managers vastly greater flexibility in managing their lands. The new, vague regulations also leave conservationists and other litigants with little opportunity to claim that an activity in a forest violates these regulations. The references to ecological conditions are quite general. Even the duty to promote ecosystem diversity is counterbalanced by the obligation to sustain existing "social and economic systems."

As this book goes to press, the validity of the new regulations has been challenged. A suit brought in California contends that the Forest Service violated the National Environmental Policy Act and the Endangered Species Act by failing to evaluate both the overall environmental effects and the particular effects on threatened and endangered species that these regulations will have.[67] In a 2007 ruling, a federal district court enjoined enforcement of the Forest Service regulations pending preparation of an environmental impact statement and consultation by the Forest Service with the Fish and Wildlife Service under the Endangered Species Act.[68] Meanwhile, litigation has continued under the old regulations, which continue to apply to existing forest plans until the Forest Service revises the plans under the new regulations. In addition, litigation continues to arise directly under the diversity mandate of the National Forest Management Act, which the Forest Service has no power to alter.

Wild species on national forests are affected, plainly, by timber harvesting and other specific management actions in the forests. Conservation groups have brought numerous suits challenging timber management, often to protect wildlife and wildlife habitat. Some of the lawsuits have challenged proposed timber cuts as violations of provisions under the National Forest Management Act that limit the Forest Service's ability to engage in clear-cutting and other even-aged timber management efforts. Clear-cutting is permitted only when it is the "optimum method" under the circumstances;

other even-aged management methods are permissible only when they are "appropriate to meet objectives and requirements of the relevant land management plan."[69] Other suits have involved claims that the Forest Service has violated the statutory provisions that require protection "for streams, streambanks, shorelines, lakes, wetlands, and other bodies of water" and that ban harvesting when "soil, slope, or other watershed conditions" will be "irreversibly damaged" or when "water conditions or fish habitat" would be "seriously and adversely" affected. Much of this litigation is likely to continue, even under the 2005 regulations, given that the lawsuits typically challenge individual timber harvesting or management efforts, not entire forest plans.

The Bureau of Land Management

Like the Forest Service, the Bureau of Land Management (BLM) operates under a general charter that has been augmented by a multiple-use mandate and land planning requirements. The BLM's first instruction came in the 1934 Taylor Grazing Act, which required the agency to "do any and all things necessary to . . . preserve the land and its resources from destruction or unnecessary injury."[70] In 1964, Congress instructed the agency to follow a multiple-use policy, much like the one that then governed the Forest Service. A more complete charter, including a reformulation of the multiple-use mandate, came in 1976, when Congress enacted the Federal Land Policy and Management Act.[71] The Federal Land Policy and Management Act defined multiple use somewhat differently than in the 1960 Multiple-Use Sustained-Yield Act applicable to the Forest Service. In the case of the BLM, multiple use must take into account not just the present needs of the American people, but their future needs. The list of multiple uses is also longer; it includes, but is not limited to, "recreation, range, timber, minerals, watershed, wildlife and fish, and natural scenic, scientific and historical values." Also, multiple-use management must avoid impairment not just of the land's productivity, but of the quality of the environment. The Federal Land Policy and Management Act specifically obligates the BLM to "give priority to the designation and protection of areas of critical environmental concern," which include areas "where special management attention is required . . . to protect and prevent irreparable damage to important historic, cultural, or scenic values, fish and wildlife resources or other natural systems or processes."

Although the Federal Land Policy and Management Act, like the National Forest Management Act, requires extensive land planning with significant public participation, it does not contain as many detailed planning requirements as the forest act. Most significantly for wildlife, the Federal Land Policy and Management Act contains no diversity provision. The BLM must prepare resource management plans and then abide by them in all its actions, including its decisions about where, when, and how intensively grazing and timber harvesting can take place. In effect, these resource management plans allocate available forage on BLM lands between livestock and wild animals, thus affecting wildlife populations. A key issue on many BLM lands in the West is how the agency will manage wild burros and horses under the Wild Free-Roaming Horses and Burros Act of 1971.[72] This act obligates the secretary of the interior to manage feral horses and burros "in a manner that is designed to achieve and maintain a thriving natural ecological balance on the public lands." At the same time, however, horses and burros labeled "excess" can be rounded up and removed from a range whenever they interfere with the achievement of land management plans, including grazing allotments.

By and large, states retain substantial control over wildlife located on lands managed by both the Forest Service and the Bureau of Land Management, at least in the case of fish and game species. For a time, beginning in the 1930s, the Forest Service expressed an interest in displacing state fish and game laws with its own administrative structure. That approach, though, did not last. Current policy is grounded chiefly on guidance that Congress gave to both agencies in 1974. In a statute known as the Sikes Act Extension, Congress instructed the agencies to "plan, develop, maintain, and coordinate programs for the conservation and rehabilitation of wildlife, fish and game."[73] The programs the agencies developed had to include "specific habitat improvement projects and related activities and adequate protection for species of fish, wildlife, and plants considered threatened or endangered." They were required also "to protect, conserve, and enhance wildlife, fish and game resources to the maximum extent practicable . . . consistent with any overall land use and management plans for the lands involved." Furthermore, and importantly, the agencies were instructed to undertake their planning in cooperation with state fish and game agencies. Finally, nothing in the Sikes Act Extension diminished "the authority or jurisdiction of the States with respect to the management of resident species of fish, wildlife, or game, except as otherwise provided by law."

At first glance, the language of the Sikes Act Extension seems to impose actual duties on the two agencies to protect and conserve wildlife. That is, it seems to go beyond the multiple-use mandates that merely require due consideration of wildlife needs. The act, however, contains other language that undercuts this interpretation, including a statement that the act does not limit the ability of the two agencies to manage their lands in accordance with multiple-use principles.

The basic federal-state arrangement in the Sikes Act Extension was soon incorporated into a provision of the Federal Land Policy and Management Act (FLPMA) in 1976. The rather awkwardly drafted provision, section 302(b), appears to ban both the Forest Service and the BLM from requiring federal permits to hunt and fish on their lands.[74] It states also that the FLPMA does not enlarge or diminish "the responsibility and authority of the States for management of fish and resident wildlife." Technically, the provision on enlarging or diminishing state authority is a limit only on the power that the agencies might exercise under FLPMA itself. Both agencies have other sources of authority—the Taylor Grazing Act, the National Forest Management Act, the Sikes Act Extension, and others. On the other hand, the history of this provision in the FLPMA suggests that Congress intended it as a basic statement of policy that both agencies were to defer to the states on fish and game issues.

The Federal Land Policy and Management Act section goes further to state that both agencies "may designate areas . . . where, and establish periods when, no hunting or fishing will be permitted for reasons of public safety, administration, or compliance with provisions of applicable law." Except in emergencies, the agencies can exercise this power to ban hunting or fishing only after consultation with state fish and game officials. The language authorizing the agencies to restrict hunting and fishing was the subject of discussion by members of Congress, and it remains somewhat ambiguous to this day. The key uncertainty is whether the agencies can ban hunting or fishing when they deem it inconsistent with the achievement of their multiple-use goals. Some legislators believed that the agencies could do so—that the term "administration" covered all agency land plans. Other members of Congress believed that "administration" simply meant land supervision and did not refer to the agencies' resource management goals. This ambiguity in the statute came to a head in several court cases in the 1970s dealing with a plan by the state of Alaska to authorize a wolf hunt on federal lands. A federal court in Alaska, hearing one of the challenges in 1977, concluded that the BLM possessed the power to halt the hunt if it wanted to do so.[75] In 1979 and 1980, two federal appellate courts ruled that, even if the BLM had the power to halt the hunt, it did not have to prepare an environmental impact statement to justify its decision not to do so.[76] Neither federal court reached the issue of whether the BLM actually had the power to prohibit the hunt.

In general, both the Forest Service and the BLM defer to states when it comes to managing fish and game populations and on the details of harvesting rules and licenses. On the other hand, both agencies retain control over the management of wildlife habitat. They thus have substantial control over which species will thrive and where. Federal-state consultation processes appear to work at least well enough to keep states and federal agencies from having their disputes spill over into courtrooms.

CHAPTER

11

The Endangered Species Act: Species Listing and Critical Habitat

For many people, the central element of federal wildlife law is the Endangered Species Act of 1973 (ESA).[1] This claim of legal primacy is likely true if the importance of a law is measured, not by animals protected, but instead by the controversy generated. In its basic structure, the Endangered Species Act is relatively simple to understand. The complications come in its details, which are numerous and sometimes confusing. This chapter and the next provide an overview of the act and highlight the key issues it raises. Necessarily, the overview deals with many of the topics rather quickly. Moreover, it does not consider the various state statutes—some similar to the federal law, others rather different—that also seek to protect rare species.

The place to begin is with Congress's announced purpose for the statute. People can have varied reasons for protecting rare species. It helps to know which of the purposes were most influential when interpreting a statute and making day-to-day decisions. According to the act's second section, the act was intended to "conserve" both "the ecosystems upon which endangered species and threatened species depend" and the species themselves.[2] Congress also explicitly stated that rare species possess "esthetic, ecological, educational, historical, recreational, and scientific value."

This language is important because various provisions of the ESA can be hard to apply without some understanding of what the act is supposed to accomplish.

- If the aim is simply to keep a species alive somewhere, then protecting it in a zoo might suffice. Certainly, a viable population of a species somewhere in the world is adequate.
- If the aim is only to protect genetic material, then storing genes in gene banks might be enough.
- If a key goal is for the species to remain available for recreational viewing in the wild, then a conservation plan might seek to protect (or reintroduce) the species in locations people can readily visit.
- The reference to scientific value might require that species be protected in their native habitat since only in such settings can scientists study species as functioning community members.
- The references to ecosystems and ecological value interject yet further policy considerations. The goal of conserving ecosystems on which rare species depend would seem to qualify as a secondary goal—that is, a goal that we pursue not as an ultimate end, but as an intermediate step—in this case, a step toward protecting the species themselves. On the other hand, the claim that the species have ecological value reverses the goals by suggesting that species deserve protection because of their ecological roles. If so, it is plainly inadequate to keep species alive in zoos or as frozen genetic material. Indeed, species protection in simply a few locations would be inadequate when particular species are needed to perform ecological functions in many locations, including regions of the country that the species once occupied and no longer does.
- Then there is the goal of protecting species so as to sustain basic evolutionary processes of speciation, a goal perhaps implicit in the concept of "ecological" value. Evolution is most likely to unfold when species exist in varied physical settings and as isolated populations. An endangered-species policy that took evolution seriously would promote multiple, independently viable populations of a given species in multiple locations, rather than a single population of the species, no matter how viable.

Courts rarely have occasion to consider these goals directly as they go about their work of resolving disputes. The various goals, though, lurk close to the surface, embedded in many questions arising under the act. What is a species, for instance, and should the government when identifying species give the most weight to genetics, morphology, or ethnology? What chance of extinction must exist before a species is considered "threatened" or "endangered"? When should a "distinct population segment" be listed for separate protection, and when has a species lost a "significant portion" of its range? How should we frame recovery goals? When is a species "in jeopardy"? When has it been "harmed"? These and other issues would be easier to resolve if we had clear, agreed-upon goals. In the absence of better guidance, the agencies that administer the Endangered Species Act—the Fish and Wildlife Service and, in the case of certain marine species, the National Marine Fisheries Service (more formally, the National Oceanic and Atmospheric Administration Fisheries Service, or NOAA Fisheries)—are called upon to reflect on these issues and, in conversation with the public, make decisions about them.

The ongoing controversy over wolves in the northern Rockies illustrates how different goals can lead to different management prescriptions. Wolves in the region have recovered to the point where they are securely established, at least so long as humans do not deliberately reduce their numbers. The return of the wolves to the region has had widespread ecological consequences that biological scientists view as highly beneficial. Making the resident wolf population more resilient is the fact that it is connected to a large wolf population in Canada. What, then, is the management goal of wolf recovery efforts, and how many wolves are needed before success is declared? If the aim is simply to restore a wolf presence to the region, present populations are likely enough. If the aim is a population large enough, within the United States alone, to secure the wolves' existence indefinitely, the present numbers may be too low to remove protections. Conversely, if genetic preservation is the sole aim, it is perhaps unclear why any wolves are needed on the United States side of the border; arguably, the genetic material is adequately protected in Canada. Finally, if wolves are desired chiefly because of their ecological roles (or for wildlife viewing), then higher populations are needed in the United States, along with wider dispersal, because wolves could provide benefits to many regions that remain unoccupied by them.

As the Endangered Species Act has been implemented, it serves to protect species characterized by low and (usually) declining populations. It is thus a species protection law, first and foremost, with ecological and recreational benefits at best secondary. Implicitly, species are protected for moral and aesthetic reasons and because of possible utilitarian value in the future, rather than (as sometimes suggested) out of a belief that all species are needed for landscapes to function properly in ecological terms.

OVERVIEW

The ESA is triggered by the listing of a species as either endangered or threatened. Most frequently, a listing is a result of a petition filed by one or more citizens requesting the listing and providing biological evidence that the species is at risk of extinction. The act specifies that decisions about listing are to be made using the best available scientific data available and that the listing agency cannot take account of the economic, social, and political consequences that might come from the classification.

Species that face the most severe dangers are listed as endangered; those facing less-severe dangers are listed as threatened. Although most of the act's protections apply only to listed species, some protections extend to species that are proposed for listing, and even to those that are candidates for study. Indeed, a major trend in the scope of the act over the years has been to find ways to protect species before they reach the brink of extinction and listing becomes necessary.

Once a species gains the dubious honor of being listed, it is covered by two types of statutory provisions. The first is intended to prevent the species' extinction, the second to increase its numbers to the point where it has been recovered and can be delisted.

There are two important extinction prevention provisions. The first is the act's broadly phrased prohibition against taking or harming an endangered species.[3] The prohibition applies not only to intentional and unintentional harms but also to certain types of habitat modification. Although this statutory provision is stated as an absolute prohibition on taking a listed species, various permit programs nonetheless allow citizens and agencies to engage in activities that incidentally harm species.

A second extinction prevention requirement applies to activities undertaken or permitted by the federal government.[4] No such federal action (including issuance of a permit for private action) can take place if it will either jeopardize the continued existence of the species or adversely modify the species' critical habitat. This prohibition is supplemented by an interagency consultation process. Before undertaking an action that could harm a species, a federal agency must consult with the Fish and Wildlife Service to clarify the possible harms to the species and figure out whether less-harmful options are available.

These extinction prevention provisions are intended to stop a species' downward slide. The act's recovery provisions are designed to go further, to increase the numbers and distribution of a listed species to the point at which it is no longer at risk and thus can be delisted. "Recovery," a synonym for "conservation," means improving a species' status so that it no longer requires protection under the act.[5] Typically, this goal requires increasing the population of a species, not simply eliminating current threats. Often it requires expanding the range that a species inhabits and promoting multiple, independently viable populations of the species so as to reduce the chance that an ecological disturbance affecting one population will wipe out the entire species. It is not enough, in short, to halt a species' decline, at least if a species has descended to the status of threatened or endangered.

The recovery process begins when the listing agency prepares a plan for the recovery of the species.[6] The plan, as we will see, is intended to guide other federal agencies in fulfilling their affirmative obligations to help recover listed species.

These are the act's main legal provisions: the prohibition on taking or harming species; the ban on jeopardizing the species or adversely modifying critical habitat; and the recovery plan. These provisions are supplemented by additional rules restricting imports, exports, and sales of listed species, dead or alive, including animal parts. On balance, the act's greatest success has been in slowing the tide of extinction. On the other hand, few species have recovered to the point where agencies can remove them from the lists. Delistings have mostly been species that were being harmed by direct or indirect takes (for example, by hunters or toxic chemicals), which were relatively easy to stop. Recovery has been more rare, and prospects are much dimmer, for many species that have declined due to habitat loss, particularly widespread land use practices. The act has proved largely ineffective in protecting habitat.

By reputation, the Endangered Species Act is commonly viewed as an absolute statute that elevates species protection above all competing interests. The reality is very different, particularly, again, in the case of habitat alteration. The statute's seemingly absolute prohibitions are subject to permit provisions that make them far more flexible and, as a result, provide much-weaker protection. In addition, enforcement is often spotty.

One final, key point: For more than three decades, endangered species protection has generated heated controversy. Pressures to gut the

Endangered Species Act have been nearly constant, along with (on the other side) recurring proposals to give the act more teeth and expand its coverage. Perhaps because of these conflicting pressures, the act has not been amended in twenty years. Political opposition, however, has been strong enough to keep the Fish and Wildlife Service (and the National Marine Fisheries Service) from having enough money to carry out their many duties. Most controversial have been the efforts the agencies make to list new species and to designate habitat that is critical to their recovery. Congress has repeatedly imposed tight financial limits on the money the agencies can spend for these purposes. (For a time Congress even insisted that the Fish and Wildlife Service spend literally no money on listing, including money donated for the purpose by private parties!) When budgets are tight, the Fish and Wildlife Service lists fewer species and designates less critical habitat. As this happens, year after year, the act becomes less of an obstacle to business as usual in America, while the nation's rare species continue to decline.

Tight budgets, it seems, are an enduring reality. Necessarily, agencies must make tough choices on how to spend their money. Citizen groups often chime in, challenging spending priorities. Prompted by disgruntled citizen groups, courts have repeatedly ordered the Fish and Wildlife Service to spend money on specific tasks mandated by law, even when the agency believes its limited resources would yield greater conservation benefits if applied elsewhere. Low budgets, in fact, increase the work and effectively waste resources. Agencies must regularly take time to explain why they have not accomplished more. They must regularly defend litigation that arises because they have not met deadlines: because they have not, for instance, completed required five-year reviews of listed species; designated critical habitat; or studied candidate species fast enough. The reality is that many ESA regulations exist solely to deal with funding shortfalls and to juggle competing responsibilities. When tight budgets and litigation by divergent interested groups meet an administration that is openly hostile to the ESA's objectives, the result is stalemate.

STANDARDS FOR LISTING SPECIES

A surprising portion of the ESA's legal complexity deals with the initial step in the conservation process, the step of identifying and listing species that are imperiled. Much of the controversy has surrounded three key subissues:

what is the unit of life that will be protected (that is, the "species"); how do we decide whether a species qualifies as threatened or endangered; and how do agencies allocate their limited funds for studying species when hundreds of species await their attention?

"Species" Defined, in General

The ESA defines the term "species" broadly to include not only any "subspecies of fish or wildlife or plants" but also any "distinct population segment" of a vertebrate animal.[7] To the uninitiated, the concept of species may seem unproblematic; the world is divided by nature into distinct forms of life—elephants are not butterflies. Biologists, however, understand the complexities of dividing life into categories. Evolution means that, despite their differences, elephants and butterflies share a common ancestor. As a result, genetic diversity exists between and among all organisms, even within a species. Not all humans, for example, have blue eyes. Furthermore, many species breed with other, similar species under stressful conditions. Much is at stake in deciding whether a given population of animals is or is not a distinct species, subspecies, or distinct population segment (a "listable unit"). If a small population qualifies as a separate listable unit, it can gain protection under the act. If, instead, the population is viewed as part of a more abundant group of animals, it may not qualify for protection, particularly if it is not a vertebrate and thus cannot be listed as a distinct population segment.

Given the breadth of the definition of what unit of life may be listed, serious questions about whether a particular group of animals should qualify as a "species" under the ESA have arisen infrequently. One spirited challenge came to a head in 2007 when a federal appellate court upheld a decision by the Fish and Wildlife Service (FWS) to list the Alabama sturgeon as a distinct species rather than include it as part of the more populous shovelnose sturgeon.[8] The challengers claimed that the agency erred in its taxonomic decision by paying too much attention to morphology—that is, to physical differences among the fish populations—rather than emphasizing genetics. In the challengers' view, the two groups were essentially identical in their genes. In its defense, the FWS responded that it had evaluated the Alabama sturgeon by paying attention to all relevant factors: to physiological, behavioral, and ecological factors as well as to genetics and morphology. The court upheld the agency's finding, concluding that the agency had considered the relevant scientific data and that its decision was supported by substantial evidence. The Fish and Wildlife Service had fared less well in 2006, however, in a controversy involving its decision to cease treating the western sage grouse as a distinct subspecies.[9] For many years, the agency had viewed the grouse as a subspecies, as did nearly all experts. In its judicial opinion, the court found that the FWS had failed to explain its reversal of policy adequately. Although the FWS has discretion to choose among conflicting scientific opinions, in the case of the western sage grouse, the only opinion by a taxonomist supported the separate taxonomic status as a subspecies. The agency's longstanding policy was to prefer the conclusions of taxonomists over those of wildlife ecologists. The FWS had not explained why it had deviated from this policy.

"Distinct Population Segments"

The most significant conflicts over the separateness of different "species" have dealt with vertebrate populations that are listed or considered for listing as "distinct population segments." The Fish and Wildlife Service can list a distinct population segment of a vertebrate animal under the act. The effect of this statutory provision is that distinct populations of a given species can have widely different conservation statuses—endangered, threatened, and unlisted—based on the risks they face. For example, the wolf was listed as endangered in all of the conterminous United States except in Minnesota and a small portion of Michigan where it was only threatened. The species was not listed in Alaska.

The ESA contains no definition of "distinct population segments." Under a policy adopted in 1996, designation as a distinct population segment is based on whether the population is physically discrete and on its significance to the species as a whole.[10] *Discreteness* focuses on whether the population is separated from other populations of the same taxon, either "as a consequence of physical, physiological, ecological or behavioral factors" or due to international borders. The *significance* of a population is determined by looking at various factors, including whether it persists in a unique or otherwise unusual ecological setting and whether the loss of the population would result in either a significant gap in the taxon's range or a loss of marked genetic characteristics. In the view of the Fish and Wildlife Service, a population qualifies as a distinct population segment only if it is both discrete and significant. Under this policy, various runs of a species of salmon—in different rivers, and fall versus spring runs—can be (and are) considered for conservation as distinct population segments. Widely separated wolf populations are also assessed separately for listing.

The requirement that a population be not just physically distinct but in some sense significant has been challenged by critics who contend that it is unduly restrictive and based on a factor—significance—that is nowhere mentioned in the Endangered Species Act. In an important ruling in 2007, a federal appellate court upheld the Fish and Wildlife position as a reasonable interpretation of the term "distinct," a term that could connote, the court stated, importance as well as physical separation.[11] In the court's view, the agency did not act arbitrarily or capriciously when it decided that the western gray squirrel population in Washington State was not sufficiently significant to qualify as a distinct population segment.

Another challenge to a ruling about distinct population segments arose in 2005, in litigation having to do with wolves in the northeastern United States. The region did not contain wolves—or at least had no breeding population—although there were wolves across the border in Canada that seemed poised to return. The Fish and Wildlife Service proposed to designate wolves in the Northeast as a distinct population segment but then reversed directions, deciding that it could not designate a distinct population segment in the absence of clear evidence that the species was present in a region. Instead, the agency lumped wolves in the Northeast (if any) with wolves in the Great Lakes region, labeling them all as a single distinct population segment separate from wolves in the northern West and in the Southwest. Two federal courts invalidated these actions as inconsistent with the act.[12] These decisions highlighted the confusion that arises when a species is reasonably plentiful in some areas but rare or absent in portions of its former range. The species as a whole cannot reasonably be listed as endangered; it is too numerous. But what, then, is the Fish and Wildlife Service to do with those animals that are not part of a population but nonetheless inhabit areas where they are, in fact, exceedingly rare? How should they be treated?

"Threatened" and "Endangered"

Once the appropriate listable unit is designated for study and potential listing—whether it is a full species, a subspecies, a distinct population segment of a vertebrate species, or a significant portion of a species, subspecies, or distinct population segment—the next task is to determine whether it qualifies as either threatened or endangered. This task requires the Fish and

Wildlife Service to assess the risks that the species faces. A "species" is endangered if it "is in danger of extinction throughout all or a significant portions of its range."[13] It is threatened if it "is likely to become an endangered species within the foreseeable future throughout all or a significant portion of its range."[14] Listing decisions are to be based on the best available scientific and commercial data available. As with the designation of a species, the assessment ignores all other factors, including the economic effects of the listing.[15]

In assessing the risk a species faces, the Fish and Wildlife Service is instructed in the Endangered Species Act to consider all relevant factors.[16] Section 4(a) of the act specifically mentions, as threats that the agency must consider, habitat loss, overharvesting, disease, predation, and all other "natural or manmade factors affecting its continued existence."[17] Significantly, the agency must also consider "the inadequacy of existing regulatory mechanisms" since a species might not require listing if existing laws already protect it adequately. In addition, section 4(b) of the act instructs the Fish and Wildlife Service to take into account all state, local, and even foreign government efforts to protect the species, "whether by predator control, protection of habitat and food supply, or other conservation practices."[18]

The problem created by these statutory definitions is that they conceal two distinctly different questions. The first is the question of risk, the probability that something bad may happen. Under the Endangered Species Act, the bad is the extinction of a species. Since extinction is a process rather than a calamitous event, it includes a time component: what is the probability that a species will become extinct in some period of time? Even possessed of full information, scientists can only provide a probability of extinction over a specified number of years because extinction involves randomness. Even more difficult is the second of the questions: is the risk acceptable? Is a species endangered if it faces a 1 percent chance of extinction over the next 1,000 years? What about, alternatively, a 15 percent chance over 100 years? The differences between these two alternatives is substantial. Assume that 5,000 years ago, our species adopted a policy of managing the environment to ensure that any mammal facing a 15 percent risk of extinction over 100 years would be protected. Had we done so, there is a 27 percent chance that there would be no living mammal species—and only a 4 percent chance that there would be more than 3 mammal species. On the other hand, had our species adopted the 1-percent-over-1,000-years approach, 95 percent of the 4,400 mammals that existed 5,000 years ago would probably still be present.

Choosing which of these alternatives is preferable is not a scientific question; it is an ethical or policy question. And it is a question that the Endangered Species Act does not clearly answer. Not surprisingly, studies of actual listing decisions suggest that the Fish and Wildlife Service is inconsistent in its determination of acceptability.

Because an agency's decisions receive considerable deference when they are challenged in court, critics of those decisions—both decisions to list and to refrain from listing—have often challenged them on the ground that the agency either failed to consider relevant scientific information or made particular factual determinations that were arbitrary and capricious. A court cannot order the Fish and Wildlife Service to gather more field data on a species; the agency is required to decide based on the best data *available.*[19] The agency does have to consider, however, all relevant data in existence. Its ruling can be set aside if material scientific evidence is ignored.

A rare case, in which "threatened" and "endangered" were translated into actual numbers, reached an Idaho federal court in 2005.[20] The dispute involved a plant (the slickspot peppergrass) rather than an animal, but the applicable legal standard was the same. The Fish and Wildlife Service withdrew a proposal to list the plant as endangered despite concluding that the species faced a 64 percent chance of extinction over the next 100 years. The agency decided that the species was not imminently in danger of extinction and thus was not "endangered." Similarly, the species was not presently "threatened" since a species is threatened only if it is likely to become endangered "within the foreseeable future." In the agency's view, 100 years ahead was beyond the "foreseeable future." The court hearing the challenge agreed with the Fish and Wildlife Service that "the foreseeable future" depended upon the species and the relevant facts and thus might be much longer for a plant or animal with a long life span. Nonetheless, the court observed, it "defied common sense," even for an annual or biennial plant species, to contend that a 64 percent chance of extinction within a century was not a sufficient threat to warrant listing. The court noted that numerous scientific groups defined endangered at much-lower threat levels, including one group of scientists convened by the National Marine Fisheries Service that defined it as a 1 percent chance of extinction within 100 years. The court found that the Fish and Wildlife Service determination was arbitrary and capricious, given the agency's failure to explain its reasoning clearly and the fact that the agency's decision contradicted the conclusions of its own scientists.

"Significant Portion of Range"

The definitions of both "endangered" and "threatened" refer to the possibility that a species might disappear from all "or a significant portion of its range." The disappearance of a species is understandable. But what does it mean for a species to disappear from "a significant portion of its range"? Countless species no longer occupy anything like the ranges they did two or three centuries ago yet are thriving in the ranges where they remain. Is a species that is plentiful in one location nonetheless properly listed as endangered because it is no longer present in a portion of its historic range? Does the fact that it is missing from a significant portion of its historic range mean that the entire species is sufficiently at risk of extinction to be listed?

This issue received scant attention until around 2000, when various courts began agreeing with conservation advocates that a species could be listed under the Endangered Species Act if its risked losing a significant part of its range, even if its status elsewhere was reasonably secure. A prominent ruling came on 2001, when a federal appellate court struck down a decision by the Fish and Wildlife Service not to list the flat-tailed horned lizard.[21] The agency concluded that the species was sufficiently secure on the public land within its range, so that the destruction of its habitat on private property did not render the species threatened. In the agency's view, habitat was "a significant portion of [a species'] range" only if the loss of that portion of its range threatened the species as a whole. The court rejected this interpretation on the ground that the words "extinction over . . . a significant portion of its range" became surplus and irrelevant if the risk of extinction had to exist for the species as a whole. In the court's opinion, the agency either had to list a species "if there are major geographical areas in which it is no longer viable but once was," or explain why it believed that the loss of range in a particular setting was not significant.

This decision implied that a species qualified for listing if it had lost a significant portion of its range, even if the loss occurred long ago. How such a listing might take place, though, was by no means clear. A species that flourished in a given range did not need the protections of the act, at least in that range. If the species did not exist in another range, then protection there seemed meaningless. What was the point of listing, as a "distinct population segment," a population with no members? The idea, no doubt, was to protect individual animals that might be attempting to recolonize a former range, but the Endangered Species Act as written contained no clear way to do so. Moreover, the decision revived questions about the underlying purpose of the act: was it merely to protect species from disappearing, or was it also to augment ecosystem function and health?

The issues remain alive. A majority of the courts that have decided the issue have followed the decision in the flat-tailed horned lizard case. A federal court in New Mexico reached the opposite conclusion in 2005, in upholding action taken by the Fish and Wildlife Service concerning the Rio Grande cutthroat trout.[22] The court sustained the agency's position that a portion of range is significant within the meaning of the statute only if it is "so important to the continued existence of a species that threats to the species in that area can have the effect of threatening the viability of the species as a whole." The court noted that threats to a species in a portion of its range might be better dealt with by protecting a subspecies or a distinct population segment. The court did not explain, however, what the Fish and Wildlife Service might do if the imperiled population on a range did not qualify as a subspecies or distinct population segment. Did the agency then have to stand back and allow a species to disappear on much of its range until the entire species faced extinction, or could it somehow take action before then?

In March 2007, the solicitor for the Department of the Interior issued an opinion on the meaning of "significant portion of its range."[23] The solicitor concluded that the term "range" refers only to the presently occupied rather than the historic range. He also determined that "significant" conferred substantial discretion on the Fish and Wildlife Service since it could refer to geographic size, the biological significance of the area, or any of the values to be protected by the Endangered Species Act, including the "esthetic, ecological, educational, historical, recreational, and scientific value" of rare species. The effect of the solicitor's opinion on the debate is unclear, since no litigation has examined its reasoning or assessed its consistency with the statute. It is perhaps worth noting that the solicitor's opinion would allow invertebrates and plants to be listed below the subspecies classification—effectively allowing the listing of distinct population segments despite the act's definition of "species."

Effects of Conservation Efforts

An additional issue that has made listing decisions more contentious relates to the instructions that Congress gave the Fish and Wildlife Service to consider efforts made by various governmental bodies to conserve species and habitat.[24] If existing conservation efforts are adequate to address the threats to a species, then listing can be unnecessary. The Fish and Wildlife Service addressed this possibility in 2003 when it issued policy guidance explaining how the agency will evaluate ongoing conservation efforts in making listing decisions.[25] In the view of the agency, the act obligates the Fish and Wildlife Service to evaluate not only governmental conservation efforts, but also conservation efforts by businesses, other private organizations, and individuals. Section 4(a)(1)(E) states that the Fish and Wildlife Service, when assessing the status of a species, must consider "manmade factors affecting its continued existence."[26] These factors, the agency decided, included conservation efforts that aid species as well as human actions that pose threats. In the case of measures that have already been implemented, the Fish and Wildlife Service looks at their records of success. In the case of measures not yet implemented, it estimates the certainty that they will be implemented and achieve the intended results.

Before the 2003 policy was issued, the Fish and Wildlife Service had often attempted to avoid listing a species by citing conservation efforts being taken to protect it. Courts invalidated these agency efforts when the specific conservation efforts that the agency relied upon were either proposed actions that had not been implemented, voluntary programs, or otherwise not legally enforceable. The common conclusion was that the agency could not rely on conservation measures, including the land use plans of other federal agencies, unless the measures had taken effect and were legally binding.[27] In its 2003 policy, the Fish and Wildlife Service sought to circumvent this legal interpretation. It reasserted that it could consider all conservation efforts, even if voluntary, so long as the efforts were reasonably certain to take effect and produce results.[28] A 2005 federal decision from Idaho discussed the new agency policy at length but did not pass judgment on its legality.[29]

For over a decade, the Fish and Wildlife Service has offered a program to encourage landowners to manage their lands to conserve species that are imperiled but not formally listed. The program invites landowners to enter into land conservation agreements—known as candidate conservation agreements—with the agency. The hope is that the voluntary conservation measures will protect the species so that listing is unneeded. When such agreements are effective, the Fish and Wildlife Service can then cite them in its listing process as a factor reducing the threats that a species faces. Although such agreements normally cover candidate species, the underlying idea is equally applicable to species already proposed for listing and even species already listed as threatened or endangered. Conservation agreements could be effective enough to warrant removing a species from the lists on the ground that the Endangered Species Act's protections are no longer required.

As an incentive to undertake conservation actions, the Fish and Wildlife Service often includes "assurances" that the landowner will not be required to do additional conservation if the species is eventually listed. In certain circumstances, the agency may issue a permit for the landowner to engage in activities that would otherwise violate the ban on taking and harming species. (Incidental take permits are considered below.)

THE LISTING PROCESS

The Endangered Species Act provides two procedural paths for listing species. The Fish and Wildlife Service (or, in the case of many marine species, the National Marine Fisheries Service) can take the initiative to study a species, evaluate its status, and then propose it for listing. This path was common during the early years of the act. Increasingly, however, listing results from the second path: any interested person may petition the agency to list a species and thereby start the review process.[30] This privately initiated process potentially provides multiple benefits. It gives individuals a chance to take an interest in particular species and work to protect them, thus promoting democracy. The workload of the Fish and Wildlife Service may be reduced when countless individuals, including highly trained scientists, do much of the investigative work on species and present the results of that work, free of charge, to the agency for its assessment. A further benefit of the individual petition process is that it can serve to check tendencies within the Fish and Wildlife Service to avoid listing a species for political purposes.

Despite these potential advantages, the second listing process, by citizen petition, is not an unalloyed good. The problem is that the Fish and Wildlife Service lacks the money to study all species that warrant attention. It must set priorities for its labors, even when the labors are aided by unpaid citizens. The agency has identified numerous species that merit attention and has compiled them into a list of candidate species. These species are evaluated based on a variety of factors relating to the importance of taking prompt action to protect them. Candidate species, that is, are studied based on the degree of urgency rather than in the order proposed for listing.

The Petition Process

To initiate the listing process, an individual must submit a petition that provides "substantial scientific or commercial information indicating that" listing "may be warranted."[31] Fish and Wildlife Service regulations set the bar high in terms of the detailed scientific evidence that petitions must include. It is not enough to send a letter expressing concern about the apparent decline of a favored species. "To the maximum extent practicable," the Fish and Wildlife Service must decide within ninety days of the receipt of a petition whether it presents enough information to satisfy the legal standard. If the evidence is inadequate, the agency can reject the petition. If the evidence is sufficient to indicate that listing "may be warranted," the agency must "promptly commence" a study of the species to determine whether to list it. The initial decision on the citizen petition, either that the requested action is or is not warranted, is published in the *Federal Register.* If the Fish and Wildlife Service finds a petition warranted and studies the species, it must decide whether listing is appropriate within twelve months of the date of the original petition.[32] Whatever the decision, the agency must publish a notice of it in the *Federal Register,* and the decision can be challenged in court.

Proposals to list species—whether based on a petition or the Fish and Wildlife Service's own evaluation—are published as proposed regulations in the *Federal Register.* A proposal includes a detailed summary (with source citations) of the conservation status of the species and an evaluation of the threats that it faces, based on the listing factors. The proposal invites public comment and may include information about a scheduled public hearing. The Fish and Wildlife Service also notifies the states in which the species is present and publishes a summary of the proposal in newspapers of general circulation. A listing proposal cannot take effect for at least ninety days. At the longer end, the agency is obligated either to issue the regulation listing the species or to withdraw the proposal within one year, unless an extension of up to six months is warranted to resolve substantial disagreement regarding the sufficiency or accuracy of the scientific data.[33]

"Warranted but Precluded"

As petitions mounted and candidate lists grew, the Fish and Wildlife Service realized that it was in trouble. Congress responded not by giving the agency the money needed to handle the petitions, but instead by amending the Endangered Species Act to allow the Fish and Wildlife Service to provide an excuse for failing to do so.[34] When handling a petition, the agency can decide that the listing petition seems meritorious—that is, that the listing is probably warranted—but that the agency is simply too busy with other work to proceed with the listing process. More precisely, the agency can decide that "the immediate proposal and timely promulgation" of a listing regulation is "precluded" by other pending proposals to list species and that "expeditious progress was being made" on these other proposals. This response to a citizen petition soon became known as the "warranted but precluded" finding. In the case of such findings, the Fish and Wildlife Service is obligated to review the petition at least annually and to decide again whether the agency should issue a proposed listing regulation or reissue a finding of warranted but precluded. While a species awaits further attention in this "warranted but precluded" limbo, the agency must monitor the status of the species "to prevent a significant risk" to its well-being.[35] The Endangered Species Act instructs the Fish and Wildlife Service, when necessary, to exercise the emergency powers it is delegated.

Inevitably, tight agency budgets have left many dozens of species in the warranted but precluded category, to the dismay of the citizens who petitioned for listing. Dismay has led to litigation, which has arguably worsened the situation because the Fish and Wildlife Service then must spend time defending its warranted-but-precluded findings rather than processing as many species as possible. The agency also spends time monitoring the status of such species and issuing annual findings on whether resources have become available to study them. Courts have firmly stated that the Fish and Wildlife Service, when issuing a warranted-but-precluded ruling, must explain with clarity exactly what listing work it is doing and must demonstrate that it is making "expeditious progress" on other listing proposals.[36] The agency cannot merely claim that it is working hard, nor can it point to other work obligations—even obligations imposed by court order—that do not involve listing.

According to a 2005 federal court ruling, the Fish and Wildlife Service cannot be sued for failure to monitor closely the status of a species covered by warranted-but-precluded ruling so long as the agency has in place (as it does) a program for monitoring such species generally.[37]

DESIGNATING "CRITICAL HABITAT"

A key part of the process of conserving a species (that is, bringing about recovery so that protection under the Endangered Species Act is no longer needed) is the identification and protection of the habitat it requires. Most species are in decline due to habitat loss or degradation. Conservation is possible only if habitat is restored and protected, an ecological reality that Congress clearly recognized in 1973.

The Endangered Species Act provides that the Fish and Wildlife Service must designate the "critical habitat" at the time a species is listed, "to the maximum extent prudent and determinable."[38] "Critical habitat" is defined as the specific areas occupied by the species "on which are found those physical or biological features" that are essential for its conservation and that "may require special management considerations or protection."[39] Critical habitat may also include areas that the species does not occupy if the Fish and Wildlife Service decides that those areas are essential to recovery of the species. Although the designation of critical habitat begins with biological science and the needs of the listed species, the process goes well beyond science. Section 4(b)(2) of the Endangered Species Act instructs the Fish and Wildlife Service to consider all effects of the habitat designation, including economic and social effects and impacts on national security.[40] The agency "may exclude" habitat that would otherwise be critical upon a finding that the overall benefits of excluding the area from habitat designation exceed the benefits of designation, unless the failure to designate the habitat "will result in the extinction of the species concerned."

Designation Standard

The process of designating critical habitat raises a number of difficult issues. Habitat need not be designated if it is "not determinable"—that is, if the Fish and Wildlife Service lacks sufficient scientific knowledge to make the designation.[41] The issue in such cases usually is one of resource availability rather than impossibility of performing the task. The "not determinable" explanation justifies only a one-year delay. At the end of that time, the agency must designate habitat based on the available information. Designation can also be avoided if it would not be "prudent." The Fish and Wildlife Service has used this justification when designation of habitat for a given species (done in publicly available maps) could increase illegal taking of the species by informing potential takers of the species' location (a particular problem with respect to endangered flowering plants and butterflies). Finally, the agency may withhold designation when it "would not be beneficial to the species," a rationale that is cited when habitat is already protected—for instance, when it is located in national parks, wilderness areas, or wildlife refuges—or a species is at risk for reasons not related to habitat loss.

The key factor is whether habitat is critical is whether it is "essential" for the "conservation" of a species. Conservation, to repeat, means recovery of the species to the point where protection under the act is no longer necessary. It is thus not enough to designate habitat that is sufficient to ward off immediate extinction. At times, however, it is difficult to decide whether particular habitat is essential, even when a species needs to expand its range in order to recover. If a species could expand its range in several alternative directions, or if various places could serve as locations to reintroduce the species, how can the Fish and Wildlife Service state that habitat in one place is essential when habitat in another might serve just as well? Plainly, the designation of critical habitat raises difficult scientific issues on the suitability of lands and water as habitats and the utility of protecting habitat corridors that link otherwise separate populations. Along with the science issues, however, are the difficulties in deciding whether particular habitat should be excluded based on economic considerations and other factors unrelated to the species. How should the agency make such calculations, given that the relevant factors call for normative judgments that go well beyond science and agency expertise?

For many years, the Fish and Wildlife Service went even further and largely refused to designate critical habitat on the view that the designation provided no protection for a species beyond that already provided by the section 7 consultation provisions of the Endangered Species Act (more fully considered below). Section 7 prohibits a federal agency from directly or indirectly engaging in acts that either "jeopardize the continued existence" of a species or "adversely modify" its critical habitat.[42] The Fish and Wildlife Service contended that the ban on "adverse modification" was violated only when the modification was so severe that it "jeopardized the continued existence" of the species. The no-jeopardy ban applied the moment a species was listed. Because the protection for critical habitat merely duplicated the no-jeopardy ban, the designation of critical habitat did not add to the act's restrictions beyond the restrictions that arose from the listing itself. Habitat designation thus conferred no additional benefits for the species—or so the agency contended.

This approach toward critical habitat was based on a clear misinterpretation of section 7. When citizens finally challenged the interpretation, courts (beginning in 2001) uniformly struck it down as a legal error.[43]

According to the courts, the section 7 ban on adversely modifying critical habitat applies whether or not the modification is so severe as to jeopardize the continued existence of the species. The designation of habitat, therefore, does provide additional protection beyond the no-jeopardy provision.[44] Particularly over the past decade, courts have regularly ordered the Fish and Wildlife Service to designate habitat in the manner prescribed by the ESA. The statute's deadlines are clear, and the duty to designate habitat is not discretionary.

Cost-Benefit Analyses

As courts invalidated the Reagan-era understanding of critical habitat, the Fish and Wildlife Service has had to begin preparing detailed studies of the full effects of designating critical habitat for particular species. As noted, the studies must consider all social and economic consequences. The purpose of a study is to enable the agency to make an informed decision about whether to forgo designating habitat because the overall costs of designation exceed the corresponding benefits. Several of these studies have led to litigation, giving rise to judicial decisions that highlight the key issues and challenges in preparing them.

One confusing issue arising from habitat studies has been whether the calculation of the economic costs of designating habitat for a species should also include the costs that arise from listing the species. Merely listing the species triggers an obligation for federal agencies to avoid actions that jeopardize the continue existence of the species. That limit on economic development is a cost due to the listing, not due to any habitat designation. The designation of critical habitat only triggers the second component of section 7—the ban on adversely modifying habitat. Logically, a study of the effects of habitat designation should exclude the consequences of listing. Indeed, as some courts have noted, any effort to include the economic costs of listing in the overall assessment could collide with the express instruction by Congress that listing decisions be made based solely on scientific considerations. After some uncertainty—due chiefly to the Fish and Wildlife Service's misreading of section 7—courts have increasingly taken the view that the economic study of habitat designation should not include the costs associated with listing.[45] The inclusion of such costs, however, is not a fatal error, so long as the decision whether to forgo designating habitat does not take these listing-related costs into account.

The resolution of this issue only begins to bring clarity to a process of study that is likely, in coming years, to stimulate litigation on a wide array of questions. According to a 2007 ruling by a federal district court, a study of critical habitat must take into account the benefits of designating habitat, including benefits in the form of species recovery.[46] Habitat designation could yield benefits in terms of protecting natural areas, enhancing ecosystem services, providing recreational opportunities, and adding market value to private lands surrounding protected areas. Many of these benefits can be assigned dollar values. But it is awkward to assign a dollar value to the protection of a rare species or to an improvement in its chances of survival. On the cost side, severe methodological issues arise in any attempt to calculate the economic consequences of restricted development. Development that is banned in one place will likely occur elsewhere, perhaps with economic gains that are every bit as helpful. Indeed, the protection of habitat in an expanding suburban area could well involve no net economic costs because the decline in market value of the protected land could be equaled or exceeded by rises in the values of other lands in the area. Similarly, it may be difficult to calculate the full effects of limiting farming or grazing in an area when the farm commodities being produced are in abundant supply nationally or when farmers and ranchers elsewhere could produce the commodities without harming listed species. What is the economic cost of pushing production elsewhere? In such instances, the economic effects on the owner of the designated land could differ markedly from the effects on the surrounding region or nation as a whole.

Ultimately, the Fish and Wildlife Service must decide, based on all the information, whether to designate all habitat essential for species recovery or whether to forgo designating some habitat due to excessive costs. The Endangered Species Act provides no standard to use in making this decision. Indeed, the language of the statute gives the agency considerable discretion. The Fish and Wildlife Service is instructed to designate habitat based on the best scientific data and "taking into consideration" the other factors. It thus seems necessary only that the agency consider the factors, without assigning them any particular weight. The decision to forgo habitat designation is permissive—the secretary of the interior (acting through the Fish and Wildlife Service) "may exclude any area from critical habitat"—rather than mandatory. Further, any such decision may take place only "if" the secretary determines that the benefits of exclusion exceed the benefits of habitat protection. Given this language, it seems likely that courts will defer to decisions by the Fish and Wildlife Service as long as it undertakes a study of critical habitat that pays attention to all material factors.

Politics and Petition Management

Lurking behind the scenes in the controversy over critical habitat designation is an issue unmentioned in the Endangered Species Act that courts have not discussed. A major reason why the Fish and Wildlife Service forgoes designating habitat is because designation could trigger a backlash. The fear of litigation is one concern; the longer that habitat designation is snarled up in litigation, the longer the time period before it takes effect. The backlash concern, though, is much broader, given the political clout of developers, land use groups, and local boosters. Can the Fish and Wildlife Service legitimately exercise its discretion to avoid designating habitat to soften resistance? Studies suggest that it has done so repeatedly. Under Section 4(b)(3)(D) of the Endangered Species Act, individuals can petition the Fish and Wildlife Service to designate habitat or to revise habitat.[47] The agency is obligated to handle such petitions in essentially the same way as an individual petition to list a species or change its listing status. In addition, individuals can sue to compel the agency to perform its nondiscretionary duty to designate habitat or to challenge its determination either that habitat is not determinable or that designation would not be prudent. According to a ruling by a Louisiana federal district court in 2007, the agency's failure to designate habitat is a continuing violation of law.[48] An individual harmed by the failure can sue without worry about any statute of limitations.

Numerous courts have expressed concern about the Fish and Wildlife Service's long delays in designating habitat.[49] On the other side, though, financial considerations carry practical importance. For the Fish and Wildlife Service, money spent studying and designating habitat is, for the most part, money not available to study species on the lengthy candidate list. Conservation groups themselves are sharply divided as to whether it is wise to push the agency to designate more habitat or whether money is better spent on new listings. Indeed, groups that have successfully pushed the Fish and Wildlife Service to spend considerable resources on habitat designation have drawn criticism from other groups that view their judicial victories as unhelpful to the larger conservation cause.

CHAPTER

12

The Endangered Species Act: Protections

The last chapter looked at the aims of the federal Endangered Species Act, considered how species are listed, and reviewed the processes and standards for designating critical habitat. The story continues in this chapter, which looks at the protections that species enjoy once they are listed or while they are under consideration for listing. Protection mostly comes from two different types of provisions in the act, briefly mentioned in the last chapter. The first are prohibitions intended to prevent the species from continuing to slide toward extinction. These include the section 7 prohibitions against jeopardizing a species or adversely affecting its habitat,[1] along with the more specific bans on harmful acts in section 9, including the wide-ranging ban on takings.[2] The other type of protections aim higher, at species recovery. These chiefly include provisions on recovery plans and affirmative agency duties to promote conservation.

SECTION 7
PROHIBITIONS AND PROCESSES

At the heart of the Endangered Species Act is the provision that bans all federal actions that jeopardize the continued existence of a species or that adversely modify designated critical habitat.[3] This provision is framed in an absolute manner and accounts for the Endangered Species Act's reputation as a powerful, inflexible statute. The provision figured prominently in the Supreme Court opinion from 1978 that brought the act to public attention, the case of *TVA v. Hill.*[4] The case pitted a rather ordinary-looking minnow—the snail darter—against the Tennessee Valley Authority's Tellico Dam. The dam, then nearly 90 percent complete, was designed to block the Tennessee River to create a reservoir. The reservoir would destroy the only known habitat of the snail darter, apparently leading to its extinction. In a firmly worded opinion, the Supreme Court ruled that construction of the dam would violate the act's jeopardy and adverse-modification prohibitions. Protecting the minnow was more important than completing the dam.

The ruling in *TVA v. Hill* was heralded as a great victory by many, though not all, conservationists while producing outrage in many quarters. It has gone down in legend as an example of excessive environmental zeal. Congress responded by establishing a special high-level federal committee and empowering it to grant exemptions for projects that it deemed worthy. The committee was quickly dubbed the "God Squad" for its ostensible power to decide which species would live and which would die. The assumption was that the committee, once formed, would promptly issue an exemption allowing completion of Tellico Dam. The committee, however, came to a startling conclusion after its study. It decided that the benefits of the dam were so slight—the project was such political pork—that it was a waste of money to complete the last 10 percent of the construction. Far from stopping a valuable project, the Endangered Species Act had halted a waste of money. Congress soon intervened, though, to authorize completion of the dam without regard either for the act or for the project's costs and benefits. Meanwhile, scientists roamed the region and found small populations of the snail darter elsewhere, enough to warrant reducing the species' status from endangered to threatened.

What few observers at the time of *TVA v. Hill* realized was how unusual the case was factually. Indeed, it has proved to be nearly unique. The dispute seemingly involved a single federal project that promised to eliminate the only known population of an endangered species. The far more common case involves a project that merely chips away at a species or its habitat without pushing it to the edge of extinction. In such cases, section 7 does not speak with nearly the same absolute tone. Indeed, it apparently allows actions to proceed unless an action, considered alone, imperils the existence of a species. As long applied, section 7 does not bar activities that merely push the species a bit closer to the edge, much less actions that only interfere with recovery efforts as measured by the impact on the entire listed unit.

As noted, courts in the past few years have struck down the Fish and Wildlife Service's erroneous reading of section 7(a)(2), making it clear that the section should be interpreted literally. As written, the provision contains two distinct, if overlapping, limits on federal agency action: it bars actions that jeopardize the continued existence of a species, and it bars actions that adversely modify habitat that has been designated as critical. The entire provision is typically termed the "no jeopardy" provision, but the name is not especially apt. The no-jeopardy part merely protects against actions that truly threaten a *species.* The no-adverse-modification language, on the other hand, is far more restrictive, at least when the species is protected by designated critical habitat and the habitat is expansive enough to provide room for recovery. In such a case, any noticeable adverse modification of the habitat would run afoul of the provision, without regard for whether it added to the species' peril. In short, the no-jeopardy rule keeps a species from being pushed too close to the edge of extinction; the habitat protection provision can halt activities that merely interfere with a species' recovery.

The Ban on Jeopardy

The no-jeopardy provision is an awkward one because it requires agencies and courts to draw a line on a continuum, usually under circumstances of considerable uncertainty. An endangered species is one that faces danger of extinction. Typically, a species is endangered because human activities have already imperiled its continued existence. Further human actions that harm the species are likely to push it, step-by-step, closer to extinction. At what point does an action jeopardize the continued existence of the *species* within the meaning of section 7? Since an endangered species is, by definition, facing danger of extinction, how much additional risk is required before the species moves from merely endangered to "in jeopardy"? The question, plainly, is one of degree. Science can evaluate risks and compare them, but the definition of "jeopardy"—like the definition of "endangered"—ultimately poses a normative question that calls for a policy judgment.

Regulations issued by the Fish and Wildlife Service define "jeopardize the continued existence of" a species as an action "that reasonably would be expected, directly or indirectly, to reduce appreciably the likelihood of both the survival and recovery of a listed species in the wild by reducing the reproduction, numbers, or distribution of that species."[5] Read closely, this regulatory stance is a rather surprising one. The statutory term "jeopardy" would seem to refer to a specific *level* of danger of extinction—something higher than merely endangered but short of a point where extinction is irreversible. Under this reading, an action would jeopardize a species if it pushed the species to this level of danger, no matter how small the push. An action would also jeopardize a species if the species is already in jeopardy and the action has any negative effect—again, no matter how slight. The Fish and Wildlife Service definition, in contrast, takes a far different approach. It defines jeopardy by looking to whether a specific action pushes a species "appreciably" to a point of greater danger, without regard for how great that danger is; it is the appreciable worsening of a species' plight that violates the section, not the absolute danger that the species faces. Arguably, a species could face exceedingly grave danger and an action that worsens its plight would nonetheless be lawful because the action did not make matters *appreciably* worse. Similarly, a species could slide downward as a result of many small actions, none of which would violate the jeopardy definition because no action, standing alone, would have an appreciable effect.

The Fish and Wildlife definition of jeopardy has drawn little attention from courts. An exception came in a 2007 federal appellate court ruling, *National Wildlife Federation v. National Marine Fisheries Service,* involving the operation of dams on the Columbia River.[6] The federal defendants argued that their activities did not jeopardize the continued existence of salmon because the challenged activities, considered in isolation, did not make the dire plight of the salmon "appreciably" worse. The federal court rejected the interpretation. "Under this approach," the court observed, "a listed species could be gradually destroyed, so long as each step on the path to destruction is sufficiently modest. This type of slow slide into oblivion is one of the very ills the ESA seeks to prevent." The court interpreted "jeopardy" as a particularly high degree of risk to a species. Any action that pushed the species to this level, however slight the action, violated the no-jeopardy rule. Similarly, an action ran afoul of the statute if it caused any additional harm to a species already in jeopardy. Going further, the Court concluded that the Fish and Wildlife Service regulations required the agency in a jeopardy determination to consider effects on species recovery, not just species survival.

"No Adverse Modification"

The legal concept of adverse modification under section 7 is also vague, although not to the same extent. The Fish and Wildlife Service definition of the term (still on the books, although invalidated by courts) provides that a habitat modification is "adverse" only if it "appreciably diminishes the value of critical habitat for both the survival and recovery of a listed species." More consistent with the statute is a definition that takes seriously the purpose of critical habitat, which is to promote species recovery. As one federal appellate court made clear in a 2007 ruling, *Center for Native Ecosystems v. Cables,* critical habitat is adversely modified in violation of section 7 by all "actions that adversely affect a species' recovery and the ultimate goal of delisting."[7]

On its face, section 7 seems to prohibit all modifications of critical habitat that diminish its value for recovery, no matter how slight the effect. The law, however, rarely bans activities that have minimal effects. Agencies and courts routinely interpret absolute bans so as to overlook what the law terms *de minimus* violations. It seems reasonable that the Fish and Wildlife Service might exclude habitat modifications that have only negligible effects on habitat or recovery. But how far can the agency go in limiting section 7 to actions that have reasonably major consequences? Can it, in essence, limit the ban to actions that have "significant" effects, or would such an interpretation (or redefinition) run contrary to congressional intent?

The Fish and Wildlife Service has not proposed a new definition of "adverse modification" to replace its invalid one, nor have courts had much occasion to consider the issue. The National Marine Fisheries Service took a stance on the issue in the Columbia River controversy, noted above, that led to the appellate court ruling in *National Wildlife Federation v. National Marine Fisheries Service.* In the litigation, the agency argued that adverse modification took place only if an action altered "an essential feature of the critical habitat" and if the alteration "appreciably diminishe[d] the value of the critical habitat for survival or recovery." This definition curtails the literal language of the statute in two ways—with the requirements that the adverse modification alter an essential habitat feature and that the alteration appreciably diminish recovery prospects. The appellate court did not comment on the proposed interpretation of section 7(a)(2) because, in its view, the agency clearly failed to evaluate dam operations to discern their full effects on recovery. The issue will no doubt return to court soon.

Discretionary Federal Actions

The bans in section 7 apply only to actions that are "authorized, funded, or carried out" by an agency of the federal government. Thus they do not apply to actions by states or private actors unless the federal government is involved, most commonly by issuing a permit or providing funding. Predictably, issues arise as to whether federal involvement in a particular action is sufficient to turn it into a federal action subject to section 7. Decisions have tended to turn on particular facts. In a typical case from 2006, a conservation group contended that section 7 applied when the state of Oregon issued a water pollution discharge permit under the federal Clean Water Act.[8] The Oregon program was subject to federal supervision and received some federal funding. In addition, the U.S. Environmental Protection Agency (EPA) had the power, if it chose, to invalidate the state-issued permit. In the court's view, however, this federal involvement was not enough to trigger section 7. The state-issued permit did not become a federal permit simply because the federal EPA chose not to invalidate it.

A further limitation of section 7 is that it applies only to *discretionary* actions by federal agencies. The Supreme Court adopted this position in a 2007 decision, *National Association of Homebuilders v. Defenders of Wildlife,* in which the Court held that section 7 did not apply when the EPA turned over the power to administer the Clean Water Act to the state of Arizona.[9] The Clean Water Act set forth an exact test for deciding whether a state could take over the program. According to the Supreme Court, the EPA had no discretion in the matter if a state program qualified under the terms of the federal statute. Because the EPA could not refuse to relinquish control to Arizona, section 7 was inapplicable.

When section 7 does apply, however, it appears that an agency must consider, in its jeopardy and adverse-modification assessments, the entire operation in which it is engaged. It cannot exclude from consideration those aspects that it deems nondiscretionary. This specific issue arose in the *National Wildlife Federation v. National Marine Fisheries Service* dispute, handed down not long after the Supreme Court ruling. The appellate court ruled in the case that the Fisheries Service had to consider in its section 7 analysis the full operation of the dams (although not the physical presence of the dams themselves). It could not exclude from consideration all activities in fulfillment of dam operations under other laws by arguing that these activities were separate and nondiscretionary.

The Consultation Process

When it enacted section 7, Congress was well aware that most federal agencies knew nothing about endangered species and would have trouble complying with the statute on their own. To address this problem, Congress created a consultation process. When an agency contemplates an action that might affect a species that is listed or proposed for listing, the agency (known as the "action agency") asks the Fish and Wildlife Service whether such a species might be present in the action area. If a species might be present, the action agency is obligated to undertake a biological assessment to determine whether the planned activity "is likely" to affect it. The statute requires an assessment for any action that could affect a listed species; Fish and Wildlife Service regulations significantly (and, no doubt, unlawfully) curtail this obligation by requiring a biological assessment only for action agency actions that are so significant that they trigger the requirement to prepare an environmental impact statement under the National Environmental Policy Act.[10]

Fish and Wildlife Service regulations leave the content of the biological assessment to "the discretion" of the action agency except that the assessment must consider the effects on both critical habitat and the listed species.[11] Once completed, the biological assessment is submitted to the Fish and Wildlife Service, which must determine within thirty days whether it concurs in the action agency's conclusion. If the assessment determines that the action is likely to affect listed species or critical habitat, then the action agency must engage in consultation with the Fish and Wildlife Service to determine whether the action will comply with the jeopardy and habitat modification prohibitions of section 7(a)(2). If the action may violate the prohibitions, the service is to suggest "reasonable and prudent alternatives" to the proposed action to eliminate the violation. Consultation under 7(a)(2) thus is designed not just to determine whether a proposed action will comply with the substantive standards (no jeopardy or adverse modification), but to identify, when a violation is likely, changes that might be made in the action to achieve compliance.

Consultation under section 7(a)(2) can take place in a variety of formal and informal means. It is usually done privately, in the sense that no information about the consultation is made public until a final decision is reached. The Endangered Species Act and its implementing regulations contain timetables and deadlines for the process. The deadlines are shorter and more difficult to avoid when the consultation involves a private party acting under a federal permit or using federal money. During consultation, the action agency must refrain from commencing its proposed activity—or, to be precise, it is to refrain from making "any irreversible and irretrievable commitment of resources" that might foreclose alternative courses of action that could avert violation of section 7(a)(2). If a proposed activity is likely to affect a species proposed for listing but not yet listed, or affect critical habitat that is proposed for designation and not yet final, then a different, less-formal process is used. The action agency is required to "confer with," rather than "consult," the Fish and Wildlife Service, and the ban on committing resources does not apply.

Biological Opinions

The outcome of an interagency consultation generally is an informal approval by the Fish and Wildlife Service of what the action agency plans to do, particularly when the action directly complies with section 7(a)(2) or when the action agency alters its proposal in response to FWS suggestions to bring it into compliance. When a consultation is not resolved informally, it leads to the issuance of a biological opinion (BiOp) by the Fish and Wildlife Service. A BiOp contains a conclusion on whether the proposed action complies with the substantive standard of section 7(a)(2). If appropriate, it also includes an explanation of reasonable and prudent alternatives that would achieve compliance. Finally—and as explained below—a BiOp includes a detailed "statement" authorizing the action agency to engage in activities that would incidentally "take" listed species. This incidental take statement thus authorizes the action agency to undertake actions that otherwise would violate the section 9 ban on takings.

BiOps and the review process leading up to them have produced a substantial amount of litigation. Litigants often contend that the Fish and Wildlife Service has overlooked critical information or has erred in its conclusion as to whether jeopardy or adverse modification will not occur.[12] Litigants also challenge the initial decision about the geographic or temporal scope of the planned action.[13] Further litigation has focused on the before-and-after comparison of the action area that is used to determine the effects of the proposed activity. This comparison requires, as a starting point, an "environmental baseline."[14] Actions that have taken place or been approved in separate section 7 consultations are typically included in the baseline so that their consequences are not considered in assessing the effects of the new proposed action. On the other hand, the assessment of an agency's proposed action must include consideration of connected and cumulative actions. That is, the "effects of the action" that must be assessed include all "direct and indirect effects . . . together with the effects of other activities that are interrelated or interdependent with that action" and also any "cumulative effects on the listed species or critical habitat."[15]

These issues are often raised by allegations that the action agency has artificially "segmented" a major action into smaller pieces in hopes that no single small piece will violate section 7(a)(2).[16] Similar complications arise when multiple federal agencies undertake actions that affect the same region or the same species, or when the federal actions are likely to stimulate future, nonfederal actions.[17] The 2007 decision on the effects of Columbia River dams on salmon, which struck down the agencies' requirement that the federal action must "appreciably" worsen the species' status, potentially will moot many of these issues.

Controversies have also arisen over the reasonable and prudent alternative proposed by the Fish and Wildlife Service in its BiOps.[18] Some of this litigation has questioned whether the alternative steps would in fact avoid the strictures of 7(a)(2), a challenge that focuses on the completeness of the scientific studies of the alternatives, especially of their consequences over longer time frames and on the affected species as a whole.[19] Other litigation has questioned whether the alternative steps will, in fact, be implemented.[20] Courts have held that a no-jeopardy BiOp cannot be based on an assumption that the alternatives will be implemented; that is, funding must be available, and the action agency must have committed to undertake the alternatives. Thus in a 2007 case from Oregon involving livestock grazing that degraded habitat for bulltrout and steelhead, *Oregon Natural Desert Association v. Lohn,* a federal court ruled that the National Marine Fisheries Service could not base a "no adverse modification" ruling on the existence of grazing management standards, given the history of noncompliance by grazers and the lack of any clear plan for dealing with noncompliance.[21]

Another source of litigation is changed conditions. What happens when the action agency, after receiving a green light from the Fish and Wildlife Service, encounters new facts or decides to change its planned action? FWS regulations require that consultation be reinitiated if new information arises or a planned activity is modified sufficiently to call into question the original BiOp.[22] Reinitiation is also required if a new species is listed, or new critical habitat designated, that could be affected by the proposed action. Sometimes the new information is a failure of planned conservation efforts to take place or a failure by an action agency to carry out broader land use plans. In other instances, reinitiation can be appropriate when new information about a listed species suggests that its plight has altered, for better or worse, since the original consultation. In several instances, courts have invalidated agency actions for failures to reinitiate consultation.[23] Finally, litigation has occasionally raised the claim that a listed species is not, in fact, present in the area—despite, perhaps, the presence of suitable habitat—and that the Fish and Wildlife Service has erred by concluding that a proposed action would jeopardize its existence.

Ultimate Responsibility

In the end, it is up to the action agency to decide whether its planned activity will comply with 7(a)(2); the Fish and Wildlife Service opinion on the issue is merely advisory. As a practical matter, however, agencies routinely defer to the judgment of the FWS, given the agency's greater expertise.[24] On their part, courts have generally allowed action agencies to rely on Fish and Wildlife's opinions so long as the action agency is forthcoming in explaining its activity to the FWS and complies with recommendations it receives in response.[25] Courts treat an FWS opinion as an independent agency action, which citizens can challenge directly in court. As a result, most litigation involving compliance with section 7(a)(2) is brought directly against the FWS rather than against the action agency. If an action agency receives a "no jeopardy" ruling from the FWS, a person who is unhappy needs to attack the FWS opinion directly.[26]

When challenged, decisions made by action agencies and the Fish and Wildlife Service under section 7(a)(2) receive deference from courts. They are overturned only if they are arbitrary and capricious or otherwise in violation of law. To withstand scrutiny, however, the agencies must consider all relevant factors and explain their reasoning with clarity. Numerous BiOps have been set aside by courts for failures to meet these requirements.

TAKINGS OF SPECIES AND PERMIT OPTIONS

The second major extinction prevention rules in the Endangered Species Act are found in section 9. This section protects listed species by prohibiting categories of activities. Unlike section 7, which applies only to activities by federal agencies, section 9 applies to the activities of all persons, public and private. Chief among the prohibitions is section 9(a)(1)(B), which makes it unlawful for any person to "take" even a single member of a listed species.[27] The term "take" is defined broadly in section 3(19) to include "harass, harm, pursue, hunt, shoot, wound, kill, trap, capture, or collect, or to attempt to engage in any such conduct."[28] The legislative history of the provision makes clear that Congress intended the term to apply broadly to all conduct that directly or indirectly harms listed species. Regulations under the Endangered Species Act elaborate two of the key terms: "harass" is defined to mean "an intentional or negligent act or omission which creates the likelihood of injury to wildlife by annoying it to such an extent as to significantly disrupt normal behavior patterns"; "harm" means "an act which actually kills or injures wildlife" and "may include significant habitat modification or degradation where it actually kills or injures wildlife by significantly impairing essential behavioral patterns, including breeding, feeding or sheltering."[29]

Habitat Alteration as "Taking"

The ban on taking plainly applies to hunting, trapping, or otherwise killing listed animals, whether intentional or not. It also applies to indirect killings through poisoning or other contamination. (Whether criminal liability would be triggered is another issue, taken up below.) The most difficult issues involving takings entail human activities that alter habitat in some way—by removing trees, constructing roads and buildings, polluting waterways, altering water flows, driving all-terrain vehicles off-road, and countless other activities. When do habitat modifications violate section 9? Further, can potential violators gain legal permission to undertake their activities despite the outcome?

The Fish and Wildlife Service regulation defining "harm" to include habitat modification was adopted by the agency under pressure from a federal appellate court, which concluded that the section 9 language covered such harms. The regulation was nonetheless challenged years later as inconsistent with the Endangered Species Act in a federal case. The dispute worked its way to the Supreme Court in 1995. The Court's ruling, in *Babbitt v. Sweet Home Chapter*, upheld the regulation as a reasonable interpretation of section 9.[30]

As written, the habitat alteration regulation does not prohibit all alterations that harm populations indirectly. Instead, it requires a showing that animals have actually been killed or injured. In the years immediately after *Babbitt,* courts hearing disputes under the regulation were typically willing to uphold a finding of harm based on significant alterations of occupied habitat that plainly disrupted breeding activities. Over the past decade, however, several courts have demanded evidence that particular animals have been injured, despite the fact that the regulation as written includes disrupted breeding and feeding as injury. Has an animal been injured, though, if it could move elsewhere to feed or reproduce? In making jeopardy determinations, the Fish and Wildlife Service has used habitat alteration as a measure of harm to species that are declining due to habitat loss. The habitat loss itself is considered harmful. Can a similar loss of occupied habitat, particularly breeding habitat, provide sufficient evidence of a violation of section 9, without need for the FWS to produce an actual animal carcass? Recent decisions provide no clear answer.

Threatened Species

The ban on taking applies only to species listed as endangered. Threatened species are addressed in section 4(d) of the Endangered Species Act, which gives the listing agency power to issue regulations for their conservation.[31] Regulations under 4(d) can make section 9 applicable to a threatened species in whole or in part. At present, the default position is that section 9 is fully applicable to a threatened species unless the listing agency by regulation specifically promulgates a different rule.

A dispute in the 1980s over wolf trapping in Minnesota raised a serious question about the scope of the Fish and Wildlife Service's discretion in extending or withholding protection for threatened species under section 9. Secretary of the Interior James Watt proposed to allow the hunting of the wolves, which were listed as threatened. A federal appellate court in 1985, in *Sierra Club v. Clark* struck down the proposed regulation as inconsistent with section 4(d).[32] The provision, as noted, authorizes the secretary of the interior to issue regulations for the "conservation" of the species. The definition of the term includes a statement that conservation could be promoted by "regulated taking" of a species "in the *extraordinary* case where population pressures within a given ecosystem cannot be otherwise relieved." In the court's view, the secretary, given this statutory language, could authorize takings of wolves only "in the extraordinary case" when the takings were necessary to deal with population pressures that could not otherwise be managed. On the facts, the secretary had made no showing that the wolf hunt was needed to deal with population pressures.

Despite the ruling in *Clark*, the Fish and Wildlife Service has issued regulations that diminish protections for particular threatened species. Sometimes this is done to deal with situations where full protection under section 9 would produce strong political resistance.

"Incidental Take" Statements

As noted above, biological opinions issued by the Fish and Wildlife Service that reach findings of no jeopardy can be issued with "statements" that authorize the action agency to carry out its planned activity, even though the activity will incidentally take listed species. So long as the action agency complies with the terms of the incidental take statement, it will not violate section 9. For the statement to be valid, it must include several elements: it must specify "the impact of such incidental taking on the species"; it must set forth the "reasonable and prudent measures" that the action agency can take to minimize the impact; and it must include requirements to report to the FWS on incidental takes as well as such other "terms and conditions" as are necessary to minimize the harm.[33] Litigation on such incidental take statements has tended to focus on their details. Several courts, drawing upon clear legislative history, have concluded that the statements must include a precise numerical limit on the number of protected animals that can be taken. When a precise numerical limit is not possible, the FWS must explain why that is so, or the incidental take statement will be invalidated.[34] Courts have also taken seriously the requirement that the statements include provisions to minimize the take, and to report to the Fish and Wildlife Service on the amount of actual take.

Challenges to incidental take statements typically are merged with challenges to the BiOps of which they are a part. Most often, alleged errors are framed in terms of deficiencies in the underlying BiOps. Thus in the 2007 ruling dealing with livestock grazing in bull trout and steelhead habitat (*Oregon Natural Desert Association v. Lohn*), discussed earlier in this chapter, the court invalidated an incidental take statement issue by the National Marine Fisheries Service because it failed to consider adequately whether grazing-induced habitat alteration would result in incidental takes. It also invalidated the statement issued by the Fish and Wildlife Service, which essentially authorized all takes incident to undertaking the authorized activities. In the court's view, the FWS statement had to include a numeric or nonnumeric limit on takes that could be monitored so that, if the limit were exceeded, interagency consultation could be reopened.

Habitat Conservation Plans and Incidental Takes

A second method for permitting incidental takes is provided in section 10(a)(1)(A), which authorizes permits "for scientific purposes or to enhance the propagation or survival of the affected species."[35] The provision was no doubt originally envisioned as a method for researchers to obtain permits to capture or disturb listed animals in the course of research. The provision was seized upon, however, for much-different use when criticisms of the Endangered Species Act mounted and defenders of the act looked for ways to make it more flexible. What evolved out of the statutory provision was habitat conservation planning.

After the first habitat conservation plan was prepared, Congress amended the ESA to formalize section 10(a)(1)(B).[36] This provision authorizes permits to take species when the take is "incidental to, and not the purpose of, the carrying out of an otherwise lawful activity." To gain a permit, a person must prepare a habitat conservation plan (HCP) that sets forth in detail: the impact that the proposed action will have on the species; steps that will be taken to "minimize and mitigate" such impacts; the funding that will be available to implement these steps"; alternative actions that the applicant considered and why the alternatives were rejected; and "such other measures that the Secretary may require as being necessary or appropriate for purposes of the plan."

Once prepared, a habitat conservation plan is submitted to the Fish and Wildlife Service for review and is made available to the public for comment. The FWS can approve the plan if it finds: that the taking will be incidental; that the applicant will minimize and mitigate the impacts of such taking "to the maximum extent practicable"; that the plan is adequately and securely funded; and that the taking "will not appreciably reduce the likelihood of the survival and recovery of the species in the wild." Before approving the plan, the agency can insist on receiving "other assurances" that the plan will be implemented. If the plan is approved, the FWS issues an incidental take permit.

Litigation over habitat conservation plans and incidental take permits has focused on the requirements for the plans and on the standards used by the Fish and Wildlife Service in reviewing them. Central to many controversies is whether the plan will actually reduce harm "to the maximum extent practicable" and whether the habitat conservation plan will be carried out over its full lifetime, which is sometimes in perpetuity. Particularly contentious are cases in which the HCP requires continued funding and the funding source is arguably insecure—for example, when it is dependent upon the continued sale of lots in an expanding subdivision or upon continued funding by a governmental entity that allocates money annually.

The ultimate standard for approving a habitat conservation plan includes a no-jeopardy standard that duplicates the language that the Fish and Wildlife Service included in its regulations to implement section 7(a)(2)—that is, the language discussed above that defines jeopardy in terms of "appreciably" reducing the likelihood of survival and recovery. As noted, this regulatory interpretation of the no-jeopardy test has recently been challenged in the context of section 7(a)(2) and does not seem consistent with the language of that section. In section 10, in contrast, the "appreciably reduce" language appears directly in the statute; it does not originate in any regulation.[37] For years, observers have assumed that the two statutory standards under sections 7 and 10 are identical despite the differing language. This assumption now seems questionable.

The issuance of an incidental take permit under section 10 is itself an action that requires consultation under section 7. (In this instance, the Fish and Wildlife Service essentially consults with itself.) A biological opinion is typically required since the permit authorizes harm to a species. According to at least one federal court, the biological study that leads up to the BiOp must include express consideration of how the incidental take will affect recovery of the species, not just its survival, given the reference in the statute to both survival and recovery.

Safe Harbors

A second type of permit intended to increase the ESA's flexibility is a "safe harbor" permit. Under it, a landowner can agree to implement conservation measures on her land, aiding or attracting listed species, without incurring a risk of liability under section 9 and with a commitment that the conservation measures can be undone if the landowner so desires. The program is designed to encourage measures that enhance prospects for a listed species without imposing risks on the landowner involved. So long as the landowner complies with the terms of the safe harbor agreement, no liability will be imposed for incidental takes. Section 10(a)(1)(A) permits are also issued in the course of candidate conservation agreements (CCAs), which (as noted above) are also aimed at enhancing the conservation. Landowners who commit to CCAs can obtain authority to engage in incidental takes of any species that is subsequently listed so long as they continue to comply with the terms of the CCA.

Section 10 and Section 7 Interplay

This intra-agency consultation under section 7 would not seem particularly significant except for the fact, now of considerable importance, that a section 7 consultation is governed by that section's substantive standards—that is, by the section 7 no-jeopardy standard and the no-adverse-modification standard. This arguably means that the Fish and Wildlife Service can issue an incidental take statement under section 10 only if the issuance of it complies not just with the section 10 no-jeopardy provision, but with the section 7 substantive standards. If this is true, then an incidental take permit is permissible only if the planned activity does not violate the section 7 no-jeopardy standard—which is, as just noted, possibly more protective of species than the section 10 no-jeopardy standard—and if it does not violate the ban on adversely modifying designated critical habitat. Thus, although section 10 only expressly includes a no-jeopardy standard, it might indirectly include a ban on modifying critical habitat due to the duty to consult under section 7 consultation. This result would upset the conventional wisdom in this area, which is that designated critical habitat is protected against harm from federal actions, but not from purely private actions.

The section 10 process for obtaining an incidental take permit is a public process with public comment and potentially hearings before the habitat conservation plan is approved and an incidental take permit issued. In addition, the requirement that the applicant prepare an HCP imposes a significant burden. For these reasons, private parties have sought to gain a federal connection to their project so that it would be subject to a section 7 interagency consultation (and the issuance of an incidental take statement) rather than an incidental take permit under section 10. When a private action does require a federal permit—for instance, a permit to deposit dredge or fill material in a waterway or wetland—the entire project can sometimes be swept under section 7, not just the action requiring the permit. Section 7 does requires a biological study as well as steps to minimize the incidental take. On the other hand, the minimization efforts are limited to the rather modest "reasonable and prudent measures." Also, section 7 lacks requirements that minimization be "to the maximum extent practicable" and with "adequate funding for the plan." For these reasons, a private landowner who can do so might prefer to have a planned project reviewed through a section 7 consultation rather than under the section 10 habitat conservation planning provisions.

"No Surprises"

In the case of incidental take permits issued under both section 10(a)(1)(A) (enhancement of conservation and safe harbors) and 10(a)(1)(B) (habitat conservation plans), the Fish and Wildlife Service has developed a policy of issuing "assurances" to some permit recipients. It instructs the permit recipients that, so long as they comply with the terms of their permits, they will not incur further liability under the Endangered Species Act, even in the event the listed species declines in numbers or other species on the same property are listed as threatened or endangered. This "no surprises" policy insulates landowners from most prospects of increased burdens. The policy was upheld by a federal court in 2007 in a much-watched case.[38] The court decided that the policy was consistent with section 10, given that the section's no-jeopardy provision only barred activities that appreciably reduced the chances of both survival and recovery of a species. This language, the court concluded, was consistent with a no-surprises policy that did not fully protect against interferences with species recovery. The court qualified its ruling, however, in what might prove a very significant way. It noted that the issuance of a section 10 incidental take permit also had to comply with section 7, with its more protective no-jeopardy standard and its no-adverse-modification standard. Thus, a permit with a no-surprises clause might well be found invalid because of a clash with section 7. In the same opinion, the court upheld the Fish and Wildlife Service's new permit revocation policy, which authorizes the revocation of an incidental take permit only if continuance of the permitted activities is inconsistent with the section 10 no-jeopardy provision.

AFFIRMATIVE DUTIES TO CONSERVE

In addition to statutory provisions designed to prevent listed species from becoming extinct, the Endangered Species Act also seeks to recover species so that they can be delisted. Two provisions of the act impose affirmative conservation duties on federal agencies, duties that require agencies to take beneficial action rather than merely to refrain from harmful conduct.

Conservation Programs

Section 7(a)(1) of the Endangered Species Act instructs the Interior and Commerce departments—the organizational homes of the Fish and Wildlife Service and the National Marine Fisheries Service, respectively—to "utilize" all their departmental programs to promote the purposes of the act.[39] All other units of the federal government are instructed to "carry out programs for the conservation of [listed] species" in consultation with the FWS and NMFS.[40] These statutory provisions quickly gave rise to a basic question: did section 7(a)(1) impose affirmative duties on agencies to take action to conserve species, and, if so, how did these duties mesh with other tasks agencies had to perform? Was the provision merely hortatory—encouraging agencies to take conservation seriously—or did it impose actual responsibilities on agencies that citizens could go to court to enforce?

Several rulings under section 7(a)(1) from the 1990s interpreted the provision as imposing vague, but nonetheless real obligations on agencies to develop programs to protect listed species.[41] Although agencies had discretion in how they went about fulfilling these obligations, they were required to do something. They couldn't treat species protection as if it were a secondary duty that they addressed only when it did not interfere with their primary responsibilities. In these judicial rulings, plaintiffs succeeded in using the section as a tool to challenge specific agency actions that seemed insensitive to the needs of listed species.

More-recent rulings have tended to read section 7(a)(1) more narrowly. Courts have interpreted it to require only that agencies develop conservation "programs," not that agencies give weight to species conservation in every action that they take.[42] Accordingly, litigants apparently can challenge an agency if it fails to develop programs, but cannot use section 7(a)(1) to attack an individual agency action taken in accordance with the agency programs. An example of this approach to section 7(a)(1) is a 2003 federal court ruling from Oregon, *Oregon Natural Resources Council Fund v. U.S. Army Corps of Engineers,* in which a conservation group challenged the Army Corps of Engineers over the operation of its dam on Elk Creek, a tributary of the Rogue River.[43] The dam interfered with migrations of the threatened coho salmon. To protect the salmon, the corps regularly trucked the fish around the dam. Plaintiffs alleged that the corps violated 7(a)(1) in three respects: in the way it operated the dam; in refusing to breach the dam; and by ineffectively managing its program to trap and haul the salmon. The court summarily rejected the claims. The corps clearly had a program in place to conserve threatened and endangered species, and it had developed the program in consultation with the appropriate agencies. It was not the court's task, under section 7(a)(1), to evaluate the success of this program or to decide whether another conservation program would work better.

Recovery Plans

Section 4(f) of the Endangered Species Act imposes on the Fish and Wildlife Service (or, again, on the National Marine Fisheries Service in the case of a marine species) an obligation to "develop and implement" a recovery plan for each listed species unless the preparation of such a plan would not promote conservation of the species.[44] The process of preparing the plan includes notice and an opportunity for public comment. The recovery plan must include:

- A description of the "site-specific management actions" that are needed to achieve conservation of the species
- "Objective, measurable criteria" for determining when recovery has taken place
- Estimates about how long it will take and how much it will cost to achieve full recovery and to achieve "intermediate steps" toward recovery

The purpose of a recovery plan is to provide a blueprint for efforts to achieve conservation. Sometimes action taken by the listing agency can suffice to achieve conservation, particularly when a species is largely or entirely located on wildlife refuges. Generally, however, the Fish and Wildlife Service alone cannot achieve recovery. Actions by other federal agencies, as well as action by states and private landowners, will be required. Section 4(f) requires the FWS not just to prepare recovery plans, but to implement them. Implementation, however, is an ambitious matter, particularly when it requires land acquisition and a range of actions that the FWS has no power to undertake or to compel.

Courts interpreting Section 4(f) have regularly concluded that recovery plans are not action documents in the sense of being legally binding. Instead, the plans are advisory. Citizens cannot compel enforcement of them, nor can the Fish and Wildlife Service itself require compliance by other parties. On the other hand, various courts have opined that recovery plans must be taken seriously, with at least some efforts made to implement them, unless an agency can explain why it has not done so. As one court explained in 1993, federal agencies "may not arbitrarily, for no reason or for inadequate or improper reasons, choose to remain idle."[45] Moreover, it is possible that courts might one day interpret Section 7(a)(1) to require agencies to help implement recovery plans. This statutory command to "implement" recovery plans apparently has at least some vague force.

Most of the litigation over recovery plans has involved either their completeness—does a plan include the required information—or, more ambitiously, the validity of the science underlying the plan. Various courts have instructed the Fish and Wildlife Service to rewrite recovery plans when the plans have not included "site-specific management actions," when they have failed to set forth "objective, measurable criteria" that define ultimate recovery, or when they have not included time and cost estimates for the work. Some courts have said that the "objective, measurable criteria" need to cover specifically all five of the factors that the FWS considers when listing (and delisting) a species—that is, criteria on habitat size and quality; on overutilization; on disease and predation; on the adequacy of protective regulatory mechanisms; and on "other natural and manmade factors."

Courts have been willing to rule against the Fish and Wildlife Service when recovery plans are incomplete under the statutory guidelines.[46] They have been less willing to second-guess the content, unless the weight of scientific opinion casts grave doubt on elements of the plan. Thus a few courts have found plans inadequate when the scientific community as a whole rejects the minimum viable population numbers that a plan contains, or when measures to protect habitat seem manifestly inadequate. Other courts have shown less willingness to question elements of a plan, contending, as one court put it, that "the substance of the plan is left to the discretion of the Secretary." In a 2007 ruling from Washington State, a federal court concluded that citizens could not sue for an "unreasonable delay" in implementing a conservation plan, at least when the agency had taken some steps to implement parts of it.[47] Even if the duty to implement the plan was mandatory (which the court doubted), a court could not compel action, given the lack of any statutory timetable for implementation.

Various commentators over the years have suggested that the Endangered Species Act might work better if Fish and Wildlife Service efforts were chiefly focused on recovery planning and on the implementation of recovery plans rather than on the activities that now draw the most attention—the designation of critical habitat, consultations under section 7, and efforts to craft habitat conservation plans so that landowners can engage in incidental takes of listed species. Conservation efforts today largely focus on defending species against new activities that threaten their continued existence. Recovery planning, in contrast, is a more affirmative activity that opens up wide arrays of options. Species are imperiled by existing activities as well as new ones. Changes in existing activities and practices might well prove cheaper and more politically acceptable than efforts to halt new development. Moreover, sound, well-publicized recovery plans can make clear to citizens the full range of actions needed for conservation to occur.

OTHER PROVISIONS

We close this lengthy consideration of the Endangered Species Act with brief comments on several aspects of the act that are either less important or less complex.

The "God Squad"

As noted, the aftermath of the *TVA. v. Hill* Supreme Court case included a revision of the Endangered Species Act by Congress to create a high-level committee—dubbed "the God Squad"—that had authority to authorize exemptions from section 7, including exemptions that result in the extinction of a species.[48] The committee comprises six cabinet-level officials and the governor of the state in which the proposed activity is to occur. A request for an exemption needs to comply with various procedural requirements. Ultimately, if five of the seven members of the committee (voting in person) agree, the committee can issue an exemption upon finding that: there are no reasonable and prudent alternatives to the action; the action is of regional or national significance; and the benefits of the action clearly outweigh the benefits of alternative courses of action consistent with conserving the listed species or its critical habitat. The committee is required to impose "reasonable mitigation and enhancement measures" to "minimize the adverse effects" on the listed species. The God Squad has seldom been called into session.

Experimental Populations

One of the conservation tools available to the Fish and Wildlife Service is the option of capturing members of a listed species and reintroducing them into unoccupied habitat. This can be done by the agency simply as an ordinary conservation activity without any particular fanfare. It can also be done, however, under section 10(j), which authorizes the creation of "experimental populations" of a listed species that can be reintroduced into new habitat with less protection for the animals.[49] The reduced protection is included not as an aid to conservation, but, to the contrary, as a way of softening local opposition to the reintroduction effort. A reintroduced population is treated as threatened, even if the species of which it is a part is listed as endangered. This means that a section 4(d) regulation can be issued, tailoring the ban on takings as the Fish and Wildlife Service sees fit. If the FWS designates the experimental population as not essential to the continued existence of the species (that is, if it is being reintroduced to promote recovery, not to avert extinction), then the population is treated under section 7 merely as a species proposed for listing (thus triggering a requirement to confer, not to consult), and no critical habitat is designated for it. To date, all experimental populations have been designated "nonessential."

The key requirement for an experimental population is that it must be at the time of reintroduction "wholly separate geographically" from nonexperimental populations of the same species. Thus, if an experimental populations becomes linked to a nonexperimental population, then the special legal status of the experimental population ends. According to a federal appellate court ruling from 2000 dealing with wolves in the northern Rockies, *Wyoming Farm Bureau Federation v. Babbitt,* the requirement of geographic separation applies to the population as a whole; it is not violated simply if individual animals overlap the ranges of the populations.[50]

Delisting

When conservation efforts for a species succeed, or if further information arises indicating that a species is in better condition than believed, the Fish and Wildlife Service can take steps to downgrade or entirely remove a species from the lists. The procedures and substantive standards are the same as those required to list a species.

Other Restrictions and Penalties

Among the protections for species contained in the Endangered Species Act are various provisions limiting imports and exports of species; banning takes of species "upon the high seas"; and banning possession, sale, delivery, transport, and so on of the species by any means.[51] A further provision extends a rather broad exemption to actions by Alaskan natives who take species for subsistence purposes, including animals and plants used to make "authentic native articles of handicrafts and clothing."[52] Detailed rules are also provided to cover species and species parts that are privately possessed at the time that a species is listed under the act.

Finally, the Endangered Species Act includes a variety of penalties for noncompliance, both civil and criminal. Sanctions vary, based on whether the species is threatened or endangered and on the violator's knowledge. More-substantial penalties are also imposed on people who are "engaged in the business as an importer or exporter of fish, wildlife, or plants."[53] Individuals accused of killing a listed species cannot raise, by way of justification, the claim that they were defending private property unless the defense is authorized by regulation (for instance, under section 4(d) regulations governing threatened species or experimental populations). Regulations do allow takes of animals when necessary for self-defense or defense of other people. A criminal prosecutor need not prove that the defendant who takes a listed animal knew of the protected status of the animal; ignorance of the Endangered Species Act is no excuse. On the other hand, some precedent suggests that the defendant must know what animal is being killed, so that a reasonable mistake as to the species could be a defense to a criminal prosecution, although not to a civil penalty.

CHAPTER

13

Biodiversity at the Landscape Scale

The topics covered in the past three chapters are linked by an underlying concern for wildlife conservation at large spatial scales, in large watersheds and landscapes. This chapter considers the challenge of landscape-scale biodiversity conservation directly. If wildlife is to survive in healthy, widespread populations, wildlife law must become increasingly intertwined with land use planning. The key to conserving wildlife is protecting habitat: wild creatures need habitat to live. The nation, in turn, can protect sufficient high-quality habitat only by taking coordinated conservation steps over wide areas, on private as well as on public land. This need will become increasingly important as the effects of global climate change begin to be felt.

Large-scale wildlife planning remains in its early stages, particularly on nonfederal lands. Necessarily, today's planning tools are tentative and experimental. In addition, large-scale planning has encountered stiff resistance, primarily from citizens who view planning as inconsistent with personal autonomy or as otherwise misguided. Opposition to planning reflects a cultural perspective that is, in fact, quite different from the one embodied in serious conservation thought. The conservation perspective views humans as part of a larger community of life, interwoven with it and ultimately dependent on its enduring health. The opposing view presents humans as autonomous beings, possessed of liberty and entitled to live as they see fit so long as they bother no one else. The conflict between these views of the human predicament—the individualistic and the communitarian—is quite substantial. The conflict is not going to disappear soon, and wildlife law is caught in the middle.

GOALS AND CHALLENGES

Wildlife conservation needs an overall goal to guide the work that is done. We have discussed a variety of goals in earlier chapters. If we arranged them chronologically based on when each goal gained public visibility, we'd see two distinct trends: wildlife law has expanded the number of species enjoying protection; and it has shifted from a rather exclusive concern with population declines to a belief that populations can be too high as well as too low.

Wildlife law's first goals were to elevate game populations, to allocate that game based on social criteria, and to destroy species considered pests or vermin. Since most wild species fit into neither the game nor the vermin category, they were left to fend for themselves, outside the law. Over time, sensibilities expanded to show concern for various nongame species, particularly ones that were charismatic or that exhibited pain when injured. Lawmakers, accordingly, extended the law's protections to many nongame birds, to furry mammals, and to large herbivores. In recent decades, cruelty-to-animals laws have expanded as well, reaching beyond domesticated and captive animals to include—at least in some contexts—wild animals that could also suffer from human cruelty. Moral and aesthetic concerns have combined to throw the law's protective mantle over rare species as well, including endangered insects. Meanwhile, municipalities have faced mounting problems with wildlife species—such as pigeons, starlings, white-tailed deer, and Canada geese—whose populations have risen to pest levels.

Ecological Aims

What, then, might be the goals that guide wildlife conservation in coming decades? The science of ecology is likely to inform these goals. But ecology, like all sciences, contains no built-in, normative framework for assessing the goodness and badness of alternative landscapes. It tells us how nature works, but not how we ought to live in it. How then do we decide when a landscape is in good condition physically, including the types and numbers of wild species in it?

For years, ecologists have explored the idea of ecological or ecosystem integrity as a useful measure of a landscape's biological composition. Some scientists have proposed integrity as a suitable overall goal for wildlife conservation, at least when economically feasible. Other conservationists have pushed, similarly, to "re-wild" the continent by bringing back big predators and as many other species as possible. These conservation approaches and others like them are sometimes called "compositionalist" because they focus on the desired biological composition of a particular landscape.

Alongside the various compositionalist goals are those that are termed "functionalist." They stress the primacy of the ways that a landscape functions, particularly in its nutrient flows and soil retention, hydrologic cycles, ability to cleanse water, and capacity for withstanding stress. Functionalist goals conserve large numbers of wild species, in widespread distributions, because of the ecological roles the species play. Where functionalist goals differ from the competing compositionalist goals is that they do not protect species that are functionally redundant, nor do they insist that landscape functions be performed by native species rather than exotics.

Not long before his death in 1948, renowned conservationist Aldo Leopold pieced together an overall conservation goal that he termed "land health." Leopold's goal was an early functionalist one. At its center Leopold placed the ability of a landscape to retain nutrients and to use them efficiently and repeatedly. Using the best science available and his best guesses to fill in knowledge gaps (which he knew were vast), Leopold thought it wise for people in each landscape to retain, whenever they could, characteristic types and numbers of wild species. Not all species would be functionally essential, Leopold surmised. But given our ignorance, it made sense to "save all the parts."

Leopold's ideal of land health largely died with him. He was, it seems, ahead of his colleagues, both in science and in the practical objectives of conservation. Since the 1980s variants of Leopold's goal have arisen. A new scientific subfield, for instance, now focuses on ecosystem health and on how we might best understand that ideal. This work overlaps with rising attention to ecological integrity, which by small steps has taken on a greater functionalist component. Even more recently, conservationists have begun portraying nature as a source of various "services" that humans enjoy and upon which they depend for survival. Rather quickly, the term "nature's services" turned into "ecological services," or "ecosystem services." When the United Nations in 2000 began a massive effort to assess the planet's overall condition, leaders of the assessment framed their inquiry and findings in terms of the planet's capacity to continue providing these ecosystem services.[1] Ecosystem services readily fit into the functionalist category of conservation goals. When these various services are lumped together, and then stripped of their human-slanted rhetoric (most notably the term "services"), they bear a striking resemblance to Leopold's goal of land health.

Given the diversities of American culture and the nation's political fragmentation, it seems unlikely that a single goal will guide American wildlife conservation. Many of the goals, though, will likely display similarities. To some degree, they will focus on the ecological functioning of landscapes with the aim of keeping the landscapes fertile and productive. They will view wild species as valuable because of their ecological roles, as well as for other purposes, such as aesthetics and recreation. Doubtless they will recognize that populations can be too high as well as too low. And they will include some preference for native species (that is, species that have inhabited a landscape for a few hundred years or more) over more-recent biological arrivals. As for the native species preference, however, many exotics are highly resistant to removal. Given that resistance, and also given limited agency budgets, it seems unlikely that states and agencies will devote much effort to eliminating exotics, except when they seriously endanger human activities or cause major wildlife declines.

The problem of exotics is among the significant challenges that wildlife policy will need to address to achieve its goals. A century and more ago, wildlife policy focused mostly on overharvesting as the cause of wildlife declines. Today, habitat loss and exotics are the dominant causes of species loss. Inevitably, conservation efforts must address these causes directly; it is no longer sufficient to limit the taking of individual animals. Habitat must be protected and, in many settings, restored if wildlife is to flourish.

The suitability of wildlife habitat can decline in many ways. For example, it can deteriorate if it no longer provides food, shelter, places to reproduce, or (for many species) places to raise young. It can also fail to meet the highly specific needs that many species have (for instance, "booming" grounds of prairie chickens). Species are interdependent, so one species can sometimes flourish only when other particular species do. Similarly, one species can decline when another overlapping species becomes too numerous. Many species require a variety of habits to order to live, and safe corridors between them. Other species require large, unbroken tracts, and are thus endangered by habitat fragmentation. Some species have little ability to survive in and among humans; they need habitat that is protected by buffer zones from human interference. Other species suffer from nearby humans less directly, as when human activities stimulate certain species to rise to exceedingly high levels, causing (in a sense) unfair interspecies competition.

These challenges are the stuff of wildlife conservation science. Once wildlife goals are set and scientific studies are done, it then becomes the task of policymakers to use their tools to address problems and promote goals. Wildlife law, that is, is something of a hybrid. One on side is wildlife science and, on the other, the policy objectives. Wildlife law needs to connect the science and the policy. Inevitably, the law must shift as changes take place in both our scientific understanding and the goals we embrace.

Landscape Planning

Details aside, tomorrow's wildlife law will surely incorporate an important element of landscape planning to protect and enhance wildlife habitat. Wildlife law, that is, will need to be joined with land planning so that wildlife conservation becomes one of several goals that planners consider as they sketch out ways for people to inhabit their landscapes.

When scientists evaluate landscapes, they often find that certain segments are far more valuable for wildlife than others. Often, landscapes contain biological "hot spots" where species diversity is high. Others contain places where particular species could live but do not, places that offer prospects for habitat expansion and species reintroduction. Similarly, landscapes can include features suitable for use by wildlife as corridors to migrate to better locations. Corridors could gain in importance with the onset of human-induced climate change. Unfortunately, many animal species can migrate more quickly than can the vegetative communities on which they depend. In addition, communities are unlikely to move as communities because the individual species are often dependent upon different climatic or soil conditions. Scientists suggest that many future ecosystems will have no current analogues. Species that require peculiar vegetative communities may thus face particularly grave difficulties responding to changing climate. To sum up, wildlife conservation will likely include more than measures to restore and protect habitat: it may include constructing new habitats in places that particular species have not inhabited for generations, if ever.

These efforts will need to address a fundamental predicament that is the flip side of the familiar "tragedy of the commons." This is the tragedy of fragmentation, the tragedy that comes when landscapes are fragmented among landowners and political jurisdictions so that there is insufficient managerial power at the landscape level to implement landscape-scale goals.

The chief obstacle here is the institution of private property rights in land. As an institution, private property entails a delegation of land management authority downward, from the community to the individual owner. This delegation of power is far from complete, of course. Governments at various levels retain the power to constrain what the owner does. But private property as commonly understood gives landowners vast autonomy in using their lands, unless their actions plainly harm neighbors. This entrenched cultural understanding of private property poses a particular challenge for efforts to protect wildlife habitat.

A related difficulty reflects the fact that regulatory control over land uses is also splintered among countless governments. This division of governmental control, like the division of lands among private owners, fuels the tragedy of fragmentation. It makes larger-scale coordination difficult, if not impossible.

Wildlife law will need to confront these forms of fragmentation. It is necessary to find ways that are consistent with American culture to improve coordination of land uses at large spatial scales and to motivate landowners to abide by the community's vision.

Future land plans to promote wildlife and other goals are likely to require a variety of legal tools and methods. They will certainly go well beyond a master land use plan that directs specific land uses from on high. When Aldo Leopold took up the issue of planning for wildlife in the 1940s, he concluded that conservation ultimately required a major change in prevailing attitudes toward land and land use. No plan, however sound and detailed, could do the trick alone, he asserted. Wildlife would truly flourish only when wildlife conservation became important to the minds and hearts of Americans. A successful wildlife plan would therefore be something quite different from plans of the past. It would be, he predicted, "a constantly shifting array of small moves, infinitely repeated, to give wildlife due representation in shaping of the future minds and future landscapes of America."[2]

PROMOTING VOLUNTARY COORDINATION

One way that American law can promote wildlife conservation is to provide legal tools for individuals to coordinate land uses on their own, through voluntary action. Landowners, of course, can already manage their personal lands for wildlife if they choose (although zoning rules in some places—weed ordinances, for instance—can pose obstacles). Legal change might help voluntary conservation by facilitating transactions among landowners, or between landowners and outsiders. Well-structured transactions can give landowners economic inducements to shift to wildlife-sensitive land uses and to do so in concert with their neighbors.

Conservation Easements, in General

One such tool is the conservation easement. An easement, as already explained, is a property interest that vests in its owner the legal power either to enter and use someone else's land (an affirmative easement) or to limit the ways that the landowner can use the land (a negative easement). Historically, American law has disfavored negative easements. By the end of the nineteenth century, American law strongly favored economic development and intensive land uses. Guided by this policy, lawmakers disliked landholding arrangements that limited what landowners could do in the future. It was felt that one human generation should not take steps that constrained the next generation's options. Negative easements could have that allegedly ill effect, depending on how they were written.

Beginning in the 1970s, many American states turned away from this inherited wisdom. They enacted statutes that authorized the creation and enforcement of a new type of negative easement, the conservation easement, that directly limited development. (States also authorized negative easements that protected historic buildings and other human creations.) A conservation easement divides land ownership into two parts. The landowner-grantor gives up a part of his or her right to use the land to the easement grantee. The easement grantee does not thereby gain that right to engage in the specific land use (if it did, the easement would be affirmative, not negative). Instead, the grantee can legally prevent the owner of the land subject to the easement (the "servient estate," it is termed) from using the land in ways that the easement prohibits.

Under statutes authorizing such easements, the parties to an easement have vast discretion over the actions that are covered. An easement could specify what land uses are permitted or instead what land uses are prohibited or allowed only under conditions. An easement could last for a specified period of years or forever.

Enforcement Uncertainties

Unfortunately, however, today's laws governing conservation easements contain significant uncertainties that limit their value as a wildlife conservation tool. Practical difficulties, mostly financial ones, also pose constraints. Conservationists need to be aware of these limits, most significantly the difficulties of enforcing easements effectively.

For starters, easement enforcement is effective only when a local court is willing to issue an injunction halting activities that violate an easement's terms. Many courts, however, are reluctant to issue injunctions. An injunction is what the law terms an "equitable remedy"—that is to say, a remedy that a judge has discretion in issuing. In addition, equitable remedies are subject to defenses that a landowner could use to block an injunction. Among the defenses is the claim of changed conditions: the claim that surrounding social and physical conditions have changed so much that issuance of the injunction would prove unhelpful or counterproductive. Another equitable defense is the defense of relative hardship. Under this defense, a court may refuse an injunction if the economic effect on the defendant/landowner considerably exceeds the benefit that would accrue to the plaintiff/easement holder.

Conservation easements are typically permanent in form, but the reality is that enforcement can prove increasingly difficult as time goes on. This is particularly true when local judges are inclined to favor local landowners and local economic development in conflicts with outside conservation groups. As a practical reality, holders of conservation easements are at the mercy of local judges, who could well be hostile to keeping local lands from development that in their view benefits the local public.

Further, the enforcement equation must also take account of economic realities. Policing and enforcing an easement can prove expensive. Who is to do the work, and where is the money to come from? A successful enforcement action merely protects the status quo. The litigation does not generate money for the plaintiff to pay enforcement costs, unless the easement imposes this duty on the servient estate owner. Nor does an enforcement action protect additional land—the kind of conservation work that might appeal to new donors. On the landowner's side, the market value of a servient estate could rise considerably if the owner got the conservation easement removed. The possibility of doing so can give a servient estate owner a strong incentive to resist, encroach, and spend thousands of dollars litigating an easement's validity or the range of activities it prohibits.

Along from these issues, there is the question of who holds legal power to enforce an easement. An easement might well be held by a conservation group that lacks the money or the inclination to enforce it. Alternatively, an easement could be held by a government body, such as a city or county—entities that often would rather see economic development than enforce the restriction.

Conservation easements take the form of private agreements between two parties, but the benefits they generate spread across surrounding landscapes. Taxpayers pay part of the easement's cost indirectly when an easement is donated and the donor takes a tax deduction for it, or when an easement's purchase price comes from a charitable contribution. Given the taxpayers' role and the obvious public benefit, perhaps interested members of the public should be able to enforce a conservation easement if the easement holder fails to do so. Similarly, when the benefits of an enforcement action extend to surrounding citizens, perhaps tax money could be used to cover the expenses of the people who enforce.

One piece of this messy enforcement predicament arose in a case resolved in 2006 by a Tennessee court.[3] The city of Chattanooga held a land conservation easement along South Chickamauga Creek. It covered land that Wal-Mart wanted to use as the site of a new supercenter. When the city preferred the new store to the continued use of the land for conservation, a private group and an individual citizen stepped forward to enforce the easement. The state court upheld their ability to do so, citing the state statute that authorized conservation easements and that allowed enforcement by their "holders and/or beneficiaries." The private citizen who used the land covered by the easement, the court held, was one of the easement's beneficiaries and was thus able to enforce it.

Perpetual conservation easements have been subject to recent criticism because they can effectively ban development on lands in ways that might later seem socially unwise. Conservation of a given parcel could make sense when first arranged but make less sense decades or generations later due to changed conditions. Nature itself could change: land that provided good habitat at one time could serve less well years later. Surrounding human activities could also change in ways that similarly undercut the original benefits.

Reform Possibilities

The law governing conservation easements needs to find mechanisms to address these issues. And it will likely do so in coming years, by slow steps. Enforcement mechanisms need to be reliable and low-cost. In addition, the people or entities able to enforce easements should be expanded in number to reduce the problems that arise when the public's interest ends up riding on the willingness of a single organization or a single government body to do the enforcing. This last problem is particularly worrisome, given that conservationists have to win year in and year out to keep land protected. Development interests, in contrast, have to win only once; they only have to convince one set of government leaders to let the bulldozers enter. The economics of enforcement would also seem to justify changes, perhaps in the direction of partial or full public funding of easement enforcement efforts. These and related legal issues have direct bearing on wildlife conservation.

The concerns with the perpetual nature conservation easements would be reduced if holders of easements could freely sell or trade them. Many land trusts are reluctant to sell or release easements, even at fair market value, just as art museums have long been unwilling to sell or trade art work that donors give them, for fear of discouraging other donors. In the case of land, however, this attitude is questionable, given that a parcel's conservation value may decline over time. By releasing an easement to the servient owner, a conservation group could receive money to purchase greater conservation benefits elsewhere. As the group did so, it could free up land that has become attractive for development. It seems wise to allow swaps of this type and to change existing laws (laws dealing with strings implicitly attached to certain easement gifts) to make them possible.

Restrictive Covenants

While conservation easements have gained the largest attention as a habitat conservation tool, a related body of property law can also function to protect habitat. Restrictive covenants are ubiquitous in the United States in urban and suburban areas. A covenant is a legally binding promise. In the case of a restrictive covenant on land, it is a promise to refrain from using the land in a particular way. Lots in residential subdivisions are typically subject to extensive covenants restricting a wide range of activities, often prohibiting all uses of the land except as a single-family residence. Enforcement can take one of two common forms. The power could reside in a homeowners' association. Or it could reside in the landowners individually, with each landowner able to enforce the limits on the land titles of neighbors.

Like easements, covenants can be drafted in nearly unlimited ways, in terms of what they restrict and who can enforce them. They can also be drafted to allow certain land uses with the permission of a designated body, such as a homeowners' association. A scheme of restrictive covenants, covering many parcels of land, could easily be drafted to protect wildlife habitat on these lands.

The difficulty with restrictive covenants as a conservation tool is that a landowner is bound by the covenant only if the landowner consented, or if a preceding owner of the land imposed the covenant as an enduring restriction. This consent requirement protects landowners from having their land use options curtailed without their permission. In the typical case, the owner of a large tract of land imposes covenants on the land and then divides the land into lots, selling them individually. The purchaser of each lot is thus bound because the covenants were in place when the owner bought. Once the land has been divided and the lots sold, covenants can be imposed only on lots with owners willing to agree to them. For the covenants to apply to all lands in a region, every owner would need to consent.

Beyond Unanimity

This unanimous consent rule, as noted, protects the options of individual owners. The downside of this rule is that it limits the powers of landowners in an area to take action when a majority of them—even a very high "supermajority"—support the conservation arrangement. Would it make good policy to allow a supermajority—say, three-quarters or four-fifths of all landowners—to impose a conservation scheme that binds all landowners, overriding the objections of dissenting owners? The idea probably strikes many Americans as inconsistent with basic ideas of private property rights, but the idea has precedent in American law and might well in time gain support. A key precedent comes from oil and gas law. Owners of land above a petroleum formation often can form drilling units or pooling arrangements that authorize drilling, production, and secondary recovery methods for all lands in the unit, even the lands of objecting owners. The petroleum deposit, that is, is understood as a common-pool resource that is best used when landowners act together rather than individually. Many secondary recovery operations can only be undertaken that way. Production would be halted or made more wasteful and costly if a common-pool development plan could be halted by the objections of any individual overlying landowner.

The case of oil and gas pooling rules and other similar precedents (for instance, the rule allowing landowners around a nonnavigable lake to use the entire lake surface, not just the water above their individual share) raise the possibility of new conservation laws in the future that treat wildlife or wildlife habitat as a common-pool resource and that similarly allow a supermajority of landowners in a region to take conservation steps that apply to all lands. If that were to happen, a supermajority of landowners along a stream, for instance, could arrange to impose setback requirements or vegetation control limits that applied to every parcel along the stream. Landowners would have lessened abilities to control their individual land parcels, but have greatly enhanced abilities to manage aspects of their lands that are ecologically interconnected with neighboring lands. A change in law of this type—unlikely at the moment, but perhaps not in the future—would need to give thought to governance issues beyond the obvious question of how large the supermajority would need to be. The issues are many: how would owners get together, what powers might they have, how would they make decisions, and should courts or other public bodies exercise an oversight role? In time, such questions may reside at the center of wildlife law.

Other Reforms

A revision such as this in the law of restrictive covenants provides an illustration of a broader category of possible legal changes that would enable landowners to work with neighboring owners to coordinate their land uses in mutually beneficial ways. These possible changes adjust the incidents of ownership so that the owners of their parcels relinquish some of their governing power to a new entity that they create with neighboring landowners and that is collectively controlled by all the owners. The effect is a pooling of interests for the shared good. Pooling of this type could occur in many ways. Experimentation, it would seem, is very much in order. Any new proposal will clash with inherited ideas about private property rights and landowner autonomy. At the same time, however, pooling arrangements can reduce the need for action by existing governments while giving landowners greater individual control. Such arrangements honor and invite individual initiative, thus enhancing democratic values.

As lawmakers give thought to enhancing such voluntary land conservation, they might find it appropriate to revise other components of American property law that currently make conservation difficult or risky. The aims of such changes would be to recognize nature conservation as a land use that is as worthy of protection as any other. Going further, they might seek to reward landowners in recognition of the ways private conservation aids the public. In terms of tax policy, for example, it might make sense to exempt high-quality habitat from all real-property taxes. The law of adverse possession might change so that landowners who dedicate their lands to wildlife conservation do not run the risk that outsiders could use their lands in ways that ripen into adverse claims of title. The law of landlord-tenant, to cite a further illustration, might be revised to ensure that tenants do not implicitly have the power to degrade valuable habitat unless the lease clearly states otherwise.

THE RIGHTS OF INDIVIDUAL LAND OWNERS

The topics just examined dealt with situations in which owners, individually or in combination, decided on their own to protect and enhance wildlife habitat. Voluntary action alone, however, is unlikely to achieve conservation goals. Particularly in the instance of wildlife corridors, the law will need to require individual landowners to limit their actions. What might this mean in terms of the rights of individual landowners? What legal forms do such laws take, and what forms might they take in the future?

Responsibilities and Regulations

This topic is extensive because wildlife and its habitat can benefit from a wide array of laws affecting individual landowners, including laws enacted with no consideration of wildlife. Many states and local governments have laws restricting development in wetlands and floodplains, on barrier islands, and on steep or unstable slopes. Although such laws serve human needs, they also aid wildlife by protecting habitat and thus are types of wildlife laws. Similar benefits come from zoning ordinances that limit development—for example, requirements that landowners avoid building near waterfronts or within certain distances of property boundaries. These laws can take many forms. A common land use rule, to cite one of many illustrations, requires landowners to leave a certain percentage of their land surface uncovered and unpaved. Open-space requirements of this type can benefit many wildlife forms, particularly when coupled with bans on fencing. In more-rural areas, lawmakers sometimes ban construction in riparian areas close to waterways. The reasoning behind such laws could lead lawmakers over time to protect wildlife habitat along roads or power easements or to dedicate edges of farm fields to wildlife use.

More-substantial protections for wildlife come from statutes that regulate extractive land uses such as mining. State forestry practices statutes that require management on a mixed-species basis or that prohibit clearcutting benefit many wild species, in part by curtailing the expansion of edge species that thrive on human disruptions of landscapes. The ideas behind forestry practices statutes apply similarly to grasslands and other vegetative communities that people use intensively. Future years thus may see the enactment of statutes prescribing ecological guidelines for the nation's vast grazing lands and other biotic communities.

Underlying these various statutes is the idea of best-management practices governing extractive uses of nature, such as farming, grazing, forestry, mining, and intensive recreation. Future laws could prescribe the proper or best practices for a wide variety of intensive land uses, aiding wildlife in the process. Best-management practices are used most visibly in state programs to reduce water pollution caused by land uses—what is termed "nonpoint pollution" because it does not originate from a discrete pipe or ditch. Experiences gained in that setting could encourage states to use best-management practices to protect wildlife habitat and aid wildlife populations.

In and around urban areas—and increasingly elsewhere, as large developments are constructed far from the urban fringe—land use rules often impose exactions on developers. For example, developers may pay into a public fund or transfer title to a small portion of their lands to gain necessary development permits. These exactions are intended to mitigate some of the areawide ill effects of the development. In a few states, land use regulators today consider the ill effects of large developments on wildlife habitat. Regulators could (and occasionally now do) require developers seeking needed permits to mitigate their ill effects by restoring and protecting wildlife habitat elsewhere. Alternatively, regulators could reduce the scale of a proposed development, or even halt it entirely, due to its impact on wildlife.

Land use laws that restrict landowners from physically excluding wild animals can also aid wildlife. In several prominent legal decisions, courts have upheld state and local laws that prohibit landowners from fencing out migrating animals—including, in one 2000 New York ruling, migrating rattlesnakes.[4] Property law, as noted in chapter 3, generally respects a landowner's desire to control physical entry onto his or her land; today, in contrast to early-nineteenth-century law, landowners are allowed to exclude public hunters. The right to exclude, however, has not been expanded to cover the exclusion of wild animals.[5] Similarly, although the U.S. Supreme

Court has held that landowners have a constitutionally protected right to compensation if a state action results in a continuing physical occupation of their lands, that protection does not apply to occupation of the land by legally protected wildlife.

There are also laws that curtail the rights of landowners to develop large parcels without taking into consideration the land's physical features. These, too, can operate as wildlife laws when they protect habitat. Along with laws protecting wetlands and other lands with particular ecological or physical features, there are laws that limit the density of development (for example, one building per forty acres). Some laws go further and ban essentially all development to protect open space around urban areas.

Recasting the Right to Develop

Such laws restricting land uses all have the potential to reduce the degradation of wildlife habitat. They would accomplish more, however, if development options were more carefully designed to minimize wildlife harms. A law that allows construction of one house per forty acres, for example, could provide more wildlife benefits if it went further and controlled where the house is constructed. The benefits would increase still more if owners of large tracts or if all landowners in a region were obligated to cluster their permitted structures to leave substantial expanses open.

A 1991 ruling by the highest court in New Jersey illustrates the possibilities that exist to tailor development rights in a landscape to allow development while at the same time maximizing ecological benefits.[6] The case involved a complex multigovernment scheme crafted to protect the unique ecological features of the Pinelands region of central and southern New Jersey. The scheme included important roles for federal and state governments while creating a regional planning commission with extensive land use control powers that formulated a comprehensive management plan. All county and municipal land use decisions had to comply with the regional plan. Many areas were entirely off-limits to development, based on ecological considerations. In other areas, the plan allowed less-dense development. Landowners could gain development credits (at the rate of two credits per forty acres) only if they agreed to conservation restrictions on their other lands. Furthermore, they could not use the development credits themselves; they could only sell the credits for use by landowners in places where more-intensive development was permitted. On the other side, owners of land slated for more-intense development could only engage in that development if they bought credits from other owners and transferred them for use on their lands. The system effectively transferred some of the profit from development to the landowners who could not develop and spread the economic benefits of development widely among land parcels, even while concentrating development in places where it did the least ecological damage.

Although land management schemes such as the Pinelands are difficult to formulate and implement, they offer considerable appeal. Using regional plans helps to overcome the tragedy of fragmentation. Coupling such plans with transferable development rights allows communities to protect particular lands—wildlife corridors, for example—and to concentrate development in more-sensible places, while also dealing with landowners fairly. In effect, such schemes transform the "right to develop" as it is commonly understood, turning it into something different but still quite valuable. A further refinement in the right to develop could include a corresponding duty to mitigate the ecological effects of the development. The mitigation could be payments by the developer into a mitigation bank, which uses the money to restore and protect sensitive lands elsewhere.

Details aside, the key point is that such schemes entail a significantly new understanding of the right to develop, one that is defined in ways that spread the gains from development among owners while allowing communities to protect landscapes. This step, if taken with some of the others described here, would amount to a significant redefinition of what landownership means in the United States. It would eliminate the idea that owners of all lands can develop without regard for nature, and replace it with the idea that owners only have the right to develop in ways that are consistent with public ecological goals. It would largely discard the idea of landownership as an abstract ideal, with the rights of all owners defined identically, and replace it with a legal scheme in which nature plays a role—in which private rights are tailored to the land and to the needs of resident wildlife.

From "Do No Harm" to Fair Share

A new understanding of private property would likely include another key component: new understandings of the types of land uses that amount to improper "harms." Landowners have long been obligated to avoid activities that harm either neighbors or the surrounding community. The idea of harm has evolved considerably over time along with American society.

Future evolution of the law might include treating land uses that significantly degrade wildlife habitat as illegal harm, just as dumping significant pollution into waterways or the air is now viewed.

At some point, such a redefinition of property ownership might incorporate an idea that is even less familiar today: the idea that each owner of rural land should bear a fair-share duty to accommodate wildlife. If wildlife is critical to the maintenance of healthy landscapes, and if landowners are the ones principally responsible for degrading wildlife habitats, is it out of line to ask that each landowner pitch in to help sustain wildlife? A frequent criticism of the Endangered Species Act is that it imposes potentially severe constraints on a small number of landowners. A fairer scheme might spread the burdens more widely, burdening "the many" more modestly rather than "the few" quite severely. For the idea even to get an audience, wildlife advocates would need to translate the vague idea into more concrete form.

If the United States is to achieve wildlife conservation goals that are phrased in terms of ecological functioning, or if it is to create major wildlife corridors, protect river corridors, and find room for wide-ranging predators, it will need to redefine the common understanding of landownership. It is difficult, however, to predict how this will happen, and whether landowners will or will not be compensated for the development limitations.

WILDLIFE AND WATER

Biodiversity conservation at landscapes scales will certainly need to pay attention not just to land uses, but to water flows—to hydrological cycles, to the timing of water flows, and to water quality. This is obvious for conserving aquatic life. It is also true of much terrestrial life. This subject is a substantial one, just like the last, yet it is possible to highlight briefly the ways that laws relating to water could change to accommodate wildlife goals.

Withdrawals

Many species suffer due to withdrawals of water from rivers and aquifers. These withdrawals and diversions are legally permissible only if users have water rights. In many states, they must also obtain permits. Water rights are subject to longstanding legal requirements that all uses of water be either reasonable, beneficial, or both. To varying degrees, the legal tests of reasonable use and beneficial use pay attention to the overall effects of a water use and to whether it is consistent with the public good. Appropriately, the public good evolves over time with changing circumstances, population increases, and shifting values. In recent decades, such changes have included a heightened concern for wildlife, including species once thought worthless. Changes in public values are reflected in the definitions of reasonable and beneficial use.

Such new definitions are already being applied to proposed new uses of water. For various reasons, however, states have been slower to apply these new definitions to existing water uses. They have been reluctant to challenge existing water uses that, by today's standards, are clearly not reasonable or beneficial. Instead, states prefer to compensate users whose uses need to end. Whatever direction they take, pressures are likely to mount to challenge water diversions that degrade fish runs, disrupt the functioning of waterways, or otherwise interfere with wildlife goals. Water law needs to take wildlife more seriously; wildlife law, in turn, needs to permeate the law of water allocation.

Pollution

Along with issues of water *quantity* are myriad legal issues about water *quality.* Pollution from discharges of effluents and land use practices harms a wide array of species. Water pollution from sources other than discrete pipes or ditches—what the federal Clean Water Act refers to as "nonpoint source" pollution—are significant problems. Indeed, in many states, the bulk of all water pollution is from nonpoint sources. The problem has become particularly acute because federal law for over three decades has imposed limits on most point sources while it has done little to address nonpoint pollution.

As in the case of other threats, wildlife law will need to confront water pollution. It will need to address and reverse land cover changes that increase runoff or reduce nature's capacity to cleanse water and cool it. A key tool in this regard might be an increased use of water quality standards that are prescribed, not in terms of maximum levels of specific chemical pollutants, but instead in terms of the ability of particular streams to sustain sensitive aquatic life. If a key aim of the Clean Water Act is to improve water quality enough to sustain fish, crustaceans, and amphibians—and one of the statute's goals is to ensure "fishable" waters—then why not specify water quality standards in terms of that goal? Similarly, why not measure the intensity of pollution coming from a given source, not by analyzing its chemical composition, but instead by testing it (in concentrated form) on aquatic life to see how much harm it causes? States are slowly moving in this direction.

Physical Alterations

Water quality and the timing of water flows are also affected by physical alterations of hydrologic systems. Wildlife law will need to address such changes as well, often challenging existing land uses that are indirectly degrading wildlife habitat, if not directly polluting wild creatures. One way to judge the health of rivers is in terms of their minimum stream flows and whether they have sufficient water to sustain life. The law could prescribe these minimum flows for rivers and streams. Although this work has begun, it is far from complete. Even states that prescribe minimum water flows rarely take into account with scientific rigor a wide range of native aquatic species. Once minimum flows are set, states need to find legal tools to protect them, restraining diversions and prohibiting land use changes and stream alterations that cause flows to fall too low.

Particularly contentious will be dams that block rivers, disrupt fish migrations, and otherwise harm aquatic species. Many dams generate hydropower and are subject to requirements that they renew their hydropower licenses every fifty years. Conservation advocates are challenging the relicensing of dams that cause particularly high levels of harm while generating low benefits, pressing to remove dams that do more harm than good. Others press dam owners to construct fish ladders or alter water release protocols so as to aid aquatic species. All these challenges, and the laws that relate to them, make up a part of wildlife law as broadly defined.

There have been clashes between dam builders and fish advocates literally for centuries. To a degree that most legal researchers find surprising, American law has long upheld the powers of states to require dam owners to add fish ladders, even after dams are built and even when dam owners were originally told ladders would not be required.[7] The issue was particularly contentious in the first half of the nineteenth century in New England, where factory dams linked to water wheels clashed with an established, profitable fishing industry. These clashes became more intense as dams impaired the ability of waterways to cleanse rivers of organic pollutants, thereby degrading drinking water supplies.

In recent decades, wildlife concerns have also focused on the standards for managing reservoir levels. Reservoirs managed strictly for human needs—for flood control, water supply, recreation, and the like—can distort downstream water flows needed by wildlife while substantially changing water temperatures. Wildlife law is expanding to play a role on this issue, too.

Drainage and Levees

In parts of the United States, a major problem is caused by efforts to quicken the pace of drying and to allow landowners to enter their lands sooner after rains. Individual landowners often install pipes or "tiles" beneath their lands to hasten drainage. Groups of landowners form drainage districts to deepen and straighten natural streams and to construct new ditches, all with the aim of lowering water tables and speeding the downstream flow of rainfall. These drainage systems can harm wildlife in many ways. They expand and accelerate water flows after rains, leading to flash floods and stream bank erosion, heightened silt loads, higher water temperatures, and other problems. Because storm water moves more quickly after rainfalls, less water percolates through the soil to enter waterways much later. The effect is to worsen low flows during dry periods, thereby concentrating pollutants and further increasing water temperatures—all to the detriment of many wild species. Here, again, wildlife law needs to cast its net widely. It needs to address drainage practices, both on individual land parcels and when undertaken by organized drainage districts.

A final and related problem arises from structural changes to rivers. Flood control levees affect waterways in multiple ways, aiding some species but harming many others. Waterway structures designed to aid ship transport systems have similar affects, as do dredging efforts. It simply is not possible to protect wildlife, particularly wide-ranging aquatic species, without confronting these practices in one way or another. If wildlife law is to achieve ecological goals and protect rare species, it cannot be cast narrowly.

FRONTIERS OF WILDLIFE LAW

The various topics covered in this chapter can be viewed as the frontier of wildlife law. They are topics that lawmakers are only beginning to address. We are likely to hear more about these topics in coming years.

Most of the topics concern efforts to protect wild animals and their habitat. Intermingled with these topics have been comments about governance processes at the landscape scale—such as the regional planning effort in New Jersey to protect the Pinelands. Before concluding, it is helpful to take up the governance issue more directly. It is in some ways the most important of wildlife law's frontiers, where the most important work needs to take place.

Many conservation efforts will succeed only if land and resource uses are coordinated at varying spatial scales. For example, well-designed regional planning efforts can shape and guide the voluntary conservation efforts described earlier in this chapter. Regional planning could also foster beneficial changes in the rights and limits of private landownership—fueling new understandings about the right to develop, about unlawful land use harms, and even introducing a fair-share duty to promote wildlife. Finally, there are the changes to the rights of water users and possible future rules protecting hydrologic systems from dams, drainage lines, levees, and vegetation removal.

Better Governance

These various individual changes, however, can fully benefit wildlife only if they are integrated into landscape-scale plans. A governance mechanism that can overcome the tragedy of fragmentation and coordinate land and resource uses is needed, and not just initially, when a plan is first formulated and implemented, but continuously thereafter, to take account of changed circumstances, to correct errors, and to accommodate nature's inherent dynamism. The best planning, the kind that is needed today, would take into account the full range of land and resource uses within a landscape and would serve multiple aims. Wildlife conservation, in other words, needs to be one of several aims that are used to formulate land use plans. It is not a goal that the law can foster in isolation.

In the past, states promoted wildlife by setting up specialty agencies charged with regulating hunting and fishing and by putting areas off-limits to harvesting. Work of that type brought benefits, particularly by protecting species whose numbers had been reduced by overharvesting. Today's challenges, however, are more difficult, as residences sprawl across rural landscapes, shopping and office centers pop up far from cities, transportation routes are more heavily used, and agricultural activities become more industrialized, chemical-dependent, and ecologically disruptive. Wildlife agencies cannot be sent off to do their work in isolation. Like wildlife law itself, they need to join up with other efforts.

A key task, then, for conservationists, legal scholars, land planners, and others, is to come up with new methods of governance at the landscape scale that integrate the land and resource uses in the landscape and that promote the full array of land use goals. What kinds of governance arrangements might work best, given the democratic values that Americans embrace and their particular dislike of having high-level governments control land uses? In what ways might landowners and other citizens come together to formulate plans for their shared landscapes?

In many parts of the country, rural landowners harbor distrust of outside interferences in how they use their lands. Urban and suburban owners are more accustomed to detailed regulatory regimes. But even in those areas, landowners insist on an ability to make some economically valuable use of their lands and balk at the notion that their options are limited. Citing the powerful principle of equality, many landowners complain with particular vehemence if they have fewer land use options than their neighbors, or if they buy land intending to develop it in the future, only to find out, when that time comes, that the land use rules have changed.

These cultural traits highlight a need to craft new governance methods. This challenge is a major frontier for wildlife law—and for natural resources and land use law generally. Too often, today's planning efforts are too fragmentary. Fish and game agencies typically have little ability to challenge the many activities that degrade wildlife habitats. It is little wonder when they adhere to the fish hatchery approach rather than addressing the root causes of declines. On the other hand, wildlife conservation cannot be the sole objective of managing most landscapes. There are simply too many competing goals more immediately linked to human needs. The answer, then, is unlikely to be simply expanding the powers of traditional wildlife agencies or state departments of natural resources.

Giving Wildlife a Voice

The special challenge is that wildlife conservation too easily fades in importance in the mix of public goals. It just seems less important. In addition, defenders of wildlife rarely stand to make money when their goals are achieved—not the way developers do when they receive a green light to develop. Too often, one side of the clash is represented by well-paid advocates while the other side must hope that good-hearted citizens can volunteer their time. When water in a river is inadequate to meet competing demands and powerful economic interests are squabbling over it, who is to stand up for aquatic life? Who is to stand up for wildlife on the urban fringe when landowners and developers stand to make millions? Who is to challenge the builders of levees, locks, and drainage lines?

The point is that when wildlife conservation is mixed with other goals into planning efforts in landscapes—as it needs to be—the goal can too readily dwindle in significance until it disappears entirely. Quite often, local governments are intent on promoting economic growth. How are planners to see the importance of wildlife conservation? And how are they to gain the expert knowledge required to craft effective conservation plans?

To dwell on this enigma is to see why public money may need to be used to undergird voices for wildlife conservation. Money could go either to advocacy-oriented government agencies or to publicly funded citizen advocates. Normal political processes are unlikely to do the job, given the gross mismatches of power. And it is unfair to burden individual citizens who care about wildlife with the duty of stepping forward, at their own expense, to perform what is clearly a public service. Wildlife conservation benefits everyone, whether or not they hunt and fish. Given that benefit, perhaps everyone should pay for wildlife defense.

Democracy and Expertise

New, landscape-scale governance methods are thus needed to coordinate land and resource uses to foster a number of public goals, some consistent with one another, some that plainly clash. Unless these processes involve landowners and other citizens, they are unlikely to gain support. At the same time, they need to draw on the expertise of professionals, including wildlife scientists. And somehow this needs to happen in ways that correct power imbalances and ensure that the public interests in wildlife receive due accord.

Several points might be made about the shapes and arrangements of these future governance arrangements. Governance is likely to be divided among several levels of government that differ in spatial scale so that the smaller levels nest in and are responsible to the higher levels. This arrangement has been successfully employed in the Pinelands and elsewhere. Guiding the allocation of power among the levels of governance could be some variant of the subsidiarity principle, so widely used in many settings. The idea is that power and responsibility should normally lie with the spatially smallest unit of government that is capable of using the power effectively and fulfilling the responsibilities. Smaller and more local is better. Many goals, though, simply cannot be achieved by uncoordinated actions at local levels—hence the need for larger-scale governance. Local levels are unlikely to be able to overcome the tragedy of fragmentation. They are also prone to compete among themselves for economically valuable land uses. Higher authority is thus needed to set overall goals and to hold lower levels accountable. It should set goals for wildlife habitat or wildlife populations, for example, and insist the water flows leaving a local region meet specified flow regimes and water quality. If sound goals are developed at higher levels, with effective means of enforcement, local governing bodies can achieve the goals as they see fit.

New or revised governance methods will need to make good use of professional expertise and advice. "Good use" means more than simply obtaining advice. It means acting upon the advice in forming land use plans. The challenges on this point are many. Legislators and other governing officials are prone to listen to constituents. As a result, a few complaining landowners along a river may be able to halt much-needed river protection. The problem of the squeaky wheel is endemic to democratic governance generally.

This familiar defect is typically addressed, with varying degrees of success, by turning work over to specialized bodies that are less prone to outside pressures. In unusual settings, the work of such a specialized body might go into effect without further action by legislatures. More often, plans must be approved by legislative bodies. Well-structured approval processes might succeed in focusing the legislative discussion on the common good rather than on personal complaints. Better processes could also make it difficult to amend plans based on individual concerns by limiting legislative bodies to approving or disapproving plans as a whole. Further, local land use plans can be subject to review by more technically oriented higher bodies that evaluate the plans based on biological criteria.

These scattered comments about governance and possible governance methods only hint at the complexities that lie ahead. Sound overall wildlife goals are needed, even if they are subject to recurring tinkering. Wildlife conservation needs to be integrated with other public goals, which requires reconciling conflicting public desires. Land plans crafted to achieve these goals will work only with substantial expertise. Overall, democratic governance needs to function well enough that the particular interests of the few do not derail efforts to promote the common good.

To raise these final issues is to see how far wildlife law has come since its early days of laws regulating hunting. It is to see also how far the law needs to go if a diverse array of wild species is to thrive in human-dominated landscapes. Wildlife law is entering a new era. It is becoming more ambitious and encompassing, yet, as it merges with other bodies of law, less distinct. However overall goals are phrased and whatever new governance processes are formulated, wildlife advocates will need new knowledge and skills. They will need to become land planners, working hand in hand with other specialists and conversant in the language of the common good. Only if they do will wildlife in America enjoy a bright future.

Acknowledgments

This work builds upon efforts that we began years ago to bring coherence to a body of law that seemed to lack it. Our efforts yielded in 2002 a rather massive textbook, intended strictly for use in law school classrooms, entitled Wildlife Law: Cases and Materials. We intended at that point to write this book fairly promptly, but other projects and events intervened. The preparation of this book required us to refresh and revisit all of our work, given the considerable changes in wildlife law over the past six years. These changes are soon to lead also to a new edition of the law school book.

We mention this earlier work because the debts we incurred in preparing it carry over to this volume. That book included a lengthy list of legal scholars, wildlife scientists, and others who helped us in one way or another. We reiterate here our thanks to them. To that list we would add our gratitude to Jamison Colburn and Kyle Landis-Marinello who provided expert commentary on particular chapters. And again, many thanks to the unpaid science advisor, J. Michael Scott.

Notes

The citation of legal materials follows a generally accepted format based on relatively simple abbreviations. Reported opinions handed down by courts are assembled in what are termed "case reporters," which are published in various book series based on the identity of the court. The standard citation format for an opinion begins with the name of the ruling, followed by the case reporter in which it appears, including volume and page numbers, and then the year of the ruling. If the name of the case reporter indirectly identifies the court that handed down the ruling, then the name of the court is not mentioned. If the name of the case reporter does not adequately identify the court, then the citation format also includes an abbreviation of the court name. Thus the famous U.S. Supreme Court ruling dealing with the snail darter and the Endangered Species Act is cited as follows: Tennessee Valley Authority v. Hill, 437 U.S. 153 (1978). The opinion appears in a series entitled United States Reports (abbreviated as U.S.). Because that series only includes decisions by the U.S. Supreme Court, the name of the court is not repeated in the citation. The opinion appears in volume 437 of U.S. Reports on page 153, and was handed down in 1978. Opinions are published roughly in chronological order. They are not organized in any way by topic or area of law. Obviously, indices are needed to find cases on particular subjects.

THE FEDERAL COURT SYSTEM

The federal court system has three levels of courts—the lowest is the federal district court; above that are the various circuit courts of appeals, and then the U.S. Supreme Court. The system is complicated by the existence of several specialty federal courts, including the United States Court of Federal Claims, from which appeals go to the United States Court of Appeals for the Federal Circuit and then to the U.S. Supreme Court.

Federal District Courts

Decisions by federal district courts appear in the series entitled Federal Supplement (F. Supp.), which has been extended into a second series (F. Supp. 2d). The standard citation format for district court opinions identifies the specific federal court that handed down the opinion—for instance, Stevens County v. U.S. Department of Interior, 507 F. Supp. 2d 1127 (E.D. Wash. 2007), a ruling handed down by the federal district court in the Eastern District of Washington in 2007. Some relatively recent judicial rulings are not published in the official series but are nonetheless available electronically through a computer data bank, Westlaw. They are cited as follows: Sierra Nevada Forest Protection Campaign v. Tippin, 2006 WL 2583036 (E.D. Cal.), a ruling handed down by the federal district court for the Eastern District of California in 2006. Rulings that are not officially reported cannot be used as precedent in litigation, but they are nonetheless useful for learning the legal views of courts.

Federal Appellate Courts

Rulings by the various federal appellate courts are published in a similar series named Federal Reporter, which has extended into a second and now third series. The federal appellate courts all have numbers, except for the appellate court in the District of Columbia (D.C. Cir.) and a specialty court called the Federal Circuit (Fed. Cir.). The standard citation form is thus: United States v. Oregon, 470 F.3d 809 (9th Cir. 2006), a ruling handed down by the United States Court of Appeals for the Ninth Circuit (a circuit that includes states along the West Coast).

THE STATE COURT SYSTEM

State reporters are gathered into a set of "regional reporters," published (like the F. Supp. and F. series) by Thomson/West Publishing. The regional lines are somewhat surprising (Illinois, for instance, is in the Northeast Reporter; Michigan is in the Northwest Reporter). The regional abbreviations are easy to grasp: for instance, A (Atlantic); S.E. (Southeast); P (Pacific). California and New York rulings appear in series limited to these states. Most states also publish their own case reporter series, with abbreviations that indicate the state. Citations to rulings in the regional reporters typically identify the court handing down the ruling, given that regions include multiple states. The supreme court of a given state is cited simply with an abbreviation that mentions the state name; lower-court rulings in the states, in contrast, are cited in ways that indicate the specific court involved. (The common abbreviation "App." refers to an intermediate appellate court that stands between trial courts and the state supreme court.) Full formal citations refer to both the regional reporter and the state reporter, if any. For instance: Moon v. North Idaho Farmers Ass'n., 140 Idaho 536, 96 P.3d 637 (2004), a ruling by the Idaho Supreme Court. If cited only to the regional reporter, this ruling would appear as 96 P.3d 637 (Idaho 2004), to indicate the court. Ryan v. New Mexico State Highway and Transp. Dept, 964 P.2d 149 (N.M. App. 1998), is a ruling handed down by the intermediate appellate court in New Mexico.

STATUTES

Statutes are generally compiled in much different ways. They are organized by subject matter, rather than chronologically, in what are termed "codes." Codes covering the laws of a given jurisdiction include nearly all statutes of continuing, widespread application. They do not include all laws enacted by a legislature—for instance, appropriations bills, private bills, and the wide variety of enactments that have effect only temporarily or in narrowly confined circumstances.

Federal Statutes

Federal statutes appear officially in a series called United States Code (U.S.C.). This code is divided chiefly into volumes and sections (§). Thus the principal charter of the National Park Service appears at 16 U.S.C. § 1, which is the first section of volume 16 of the U.S. Code. More often used and cited is a much longer version of the U.S. Code that includes references to judicial rulings interpreting each statutory section, along with references to other research materials. This series is known as United States Code Annotated (U.S.C.A.) and is cited similarly. Many volumes are quite extensive and take up multiple physical books. The volume and section numbers are the same in this commercial book series as in the United States Code. Some well-known statutes appear all in one place in the code, in consecutive statutory sections. Thus the Endangered Species Act appears at 16 U.S.C. §§ 1531-1544. (The version that appears there includes all revisions to the act made since it was originally enacted in 1973. The original 1973 version of the act can be found in another series, U.S. Statutes at Large, which publishes federal statutes in chronological order, indicated by Public Law numbers.) Other well-known statutes are added to the United States Code not as a cohesive whole, in consecutive new sections to the code, but instead as revisions to various existing sections and as new sections added among the sections of the existing code.

State Statutes

All states have their own, somewhat similar codes, sometimes published by the state government, sometimes by an authorized commercial publishing company. The names of the codes vary—as "codes," "compiled statutes," "revised statutes," and the like. Thus the code in Arizona is termed Arizona Revised Statutes (A.R.S.). Tennessee statutes appear in Tennessee Code Annotated (Tenn. Code Ann.). Any citations to a statutory section is presumed to refer to the section as then in effect. Dates given for statutory sections in state codes do *not* refer to the date when the specific statute was enacted; instead, they refer to the year in which the book series was published. Specific mention is made if the statute has been amended since the publication of the book series.

FEDERAL AND STATE CONSTITUTIONS

Citations to federal and state constitutional provisions are typically made directly to the particular constitution, with no indication of any book series in which it appears. Most constitutions are organized into articles and specific sections, along with amendments. Thus the provision of the Montana constitution stating that citizens have a right to a clean and healthful environment is cited simply as Mont. Const. Art. 2, § 3.

OLDER LEGAL PRECEDENTS

Legal precedents prior to approximately 1870 typically appear in different formats and are more difficult to decipher without reference to a guide to legal citations. Many early judicial opinions were published in volumes that carried the surname of the person designated as judicial reporter. Statutes no longer in effect are similarly cited to an array of original reporters. English statutes no longer in effect are cited to a series linked to the ruling king or queen and the year of the reign. These statutes are often given titles, typically based on the opening words of the statute. Thus a sixteenth-century statute imposing a closed season on waterfowl is cited as To Avoid Destroying Water-Fowl, 25 Hen. 8, ch. 11, § 2 (1533), which appeared in the twenty-fifth year of the reign of King Henry VIII, as chapter 11, section 2 of the volume. (It seems likely that Henry's formal break with the Catholic Church and the birth of Elizabeth, later queen, were bigger news stories that year.)

Chapter 1

1. Bald and Golden Eagle Protection Act, 16 U.S.C. § 668 note.
2. Colo. Const. Art. 18, § 12(b).
3. State v. Buford, 331 P.2d 1110 (N.M. 1958).
4. State v. Cleve, 980 P.2d 23 (N.M. 1999).
5. N.M. Laws 1999, ch. 107, effective July 1, 1999, codified at N.M.S.A. § 30-18-1.
6. N.M.S.A. § 30-18-1 (A) (1).
7. N.M. Laws 2007, ch. 6, § 1, effective June 15, 2007.
8. Boushehry v. State, 648 N.E.2d 1174 (Ind. App. 1995).
9. Koop v. United States, 296 F.2d 53 (8th Cir. 1961).
10. Migratory Bird Treaty Act, 16 U.S.C. §§ 703-711.
11. Schultz v. Morgan Sash & Door Co., 344 P.2d 253 (Okla. 1959).
12. State v. Couch, 341 Or. 610, 147 P.3d 322 (2006).
13. Key v. State, 215 Tenn. 136, 384 S.W.2d 22 (1964).
14. Federal Aid in Wildlife Restoration Act [Pittman-Robertson Act], 16 U.S.C. §§ 669-669i; 26 U.S.C. §§ 4161, 4181.

15. Federal Aid in Fish Restoration Act [Dingell-Johnson Act], 16 U.S.C. §§ 777-777l; 26 U.S.C. § 9504 (a).
16. Land and Water Conservation Fund Act, 16 U.S.C. §§ 460l to 460l-11.

Chapter 2

1. Paul Shepard, *The Others: How Animals Made Us Human* (Washington, DC: Island Press, 1996), p. 5.
2. Pa. Const. Art. I, 2.7.
3. Mont. Const. Art. 2, § 3.
4. Mass. Const. Amend. Art. 49.
5. Minn. Const. Art. 13, § 2.
6. Va. Const. Art. 11, § 4.
7. Blades v. Higgs, 11 House Lords Cases 621, 11 English Reports 1474 (House of Lords 1865).
8. Cawsey v. Brickey, 144 P. 938 (Wash. 1914).
9. State v. Rodman, 59 N.W. 1098 (Minn. 1894).
10. *Ex parte* Maier, 37 P. 402 (Cal. 1894).
11. State v. McHugh, 630 So.2d 1259, 1265 (La. 1994).
12. Owsichek v. State, 763 P.2d 488, 495 (Alaska 1988).
13. Ala. Code, § 9-11-230.
14. Ariz. Rev. Stat. § 17-103.
15. Colo. Rev. Stat. § 33-1-101.
16. Wyo. Stat. § 23-1-103.
17. Geer v. Connecticut, 161 U.S. 519 (1896).
18. Missouri v. Holland, 252 U.S. 416 (1920).
19. Hughes v. Oklahoma, 441 U.S. 322 (1979).
20. O'Brien v. State, 711 P.2d 1144 (Wyo. 1986).
21. Pullen v. Ulmer, 923 P.2d 54, 60 (Alaska 1996).
22. State v. Fertterer, 841 P.2d 467 (Mont. 1992).
23. Barrett v. State, 116 N.E. 99 (N.Y. 1917).

Chapter 3

1. Pierson v. Post, 3 Caines 175 (N.Y. Sup. Ct. 1805).
2. Bethany R. Berger, "It's Not About the Fox: The Untold Story of *Pierson v. Post*." 55 *Duke Law Journal* 1089 (2006).
3. Ghen v. Rich, 8 F. 159 (D. Mass. 1881).
4. Liesner v. Wanie, 145 N.W. 374 (Wis. 1914).
5. State v. Shaw, 65 N.E. 875 (Ohio 1902).
6. Ferguson v. Miller, 1 Cow. 243 (N.Y. Sup. Ct. 1823).
7. People v. Sanders, 696 N.E.2d 1144 (Ill. 1998).

8. Shuger v. State, 859 N.E.2d 1226 (Ind. App. 2007).
9. State v. Ball, 796 A.2d 542 (Conn. 2002).
10. Commonwealth v. Haagensen, 900 A.2d 468 (Pa. Cmnwlth. 2006).
11. M'Conico v. Singleton, 9 S.C.L. (2 Mill.) 244 (1818).
12. Frame of Government of Pennsylvania § XXII (1683), reprinted in *Sources and Documents of United States Constitutions,* ed. William F. Swinder, vol. 8 (Dobbs Ferry, NY: Oceana Publications, 1979), 263, 266.
13. Vt. Const. of 1793/6 ch. II, § 40, reprinted in *Sources and Documents of United States Constitutions,* ed. William F. Swinder, vol. 9 (Dobbs Ferry, NY: Oceana Publications, 1979), 507, 514.
14. John Woods, *Two Years' Residence on the English Prairie of Illinois* (Chicago: R. R. Donnelley, 1968; orig. pub. 1822).
15. John Mack Faragher, *Sugar Creek: Life on the Illinois Prairie* (New Haven: Yale University Press, 1988), p. 132.
16. William Elliott, *Carolina Sports by Land and Water, Including the Incidents of Devil-Fishing,* facsimile of original 1846 edition (New York: Arno Press, 1967), 166–72.
17. Hunters, Anglers and Trappers Association of Vermont, Inc. v. Winooski Valley Park District, 913 A.2d 391 (Vt. 2006).
18. Rosenthal-Brown Fur Co. v. Jones-Frere Fur Co., 110 So. 630 (La. 1926).
19. James v. Wood, 19 A. 160 (Me. 1889).
20. Bilida v. McCleod, 211 F.3d 166 (1st Cir. 2000).
21. State ex rel. Visser v. State Fish & Game Commission, 437 P.2d 373 (Mont. 1968).
22. Goff v. Kilts, 15 Wend. 550 (N.Y. Sup. Ct. 1836).
23. Brown v. Eckes, 160 N.Y.S. 489 (City Ct. Yonkers, 1916).
24. Mullett v. Bradley, 53 N.Y.S. 781 (N.Y. App. Div. 1898).
25. State v. House, 65 N.C. 315 (1871).
26. Haywood v. State, 41 Ark. 479 (1883).
27. Andrus v. Allard, 444 U.S. 51 (1979).
28. 16 U.S.C. § 668-668d. The act was amended in 1962 to add golden eagles, and the statute was renamed the Bald and Golden Eagle Protection Act.
29. Graves v Dunlap, 152 P. 532 (Wash. 1915).
30. State v. Brogan, 862 P.2d 19 (Mont. 1993).
31. Sollers v. Sollers, 26 A. 188 (Md. 1893).
32. E. A. Stephens & Co. v. Albers, 256 P. 15 (Colo. 1927).
33. State v. Bartee, 894 S.W.2d 34 (Tex. App., San Antonio, 1994).

Chapter 4

1. Cawsey v. Brickey, 144 P. 938 (Wash. 1914).
2. Harper v. Galloway, 51 So. 226 (Fla. 1910).
3. State v. Herwig, 117 N.W.2d 335, 2337 (Wis. 1962).

4. Collopy v. Wildlife Commission, 625 P.2d 994 (Colo. 1981).
5. Clayjon Production Corp. v. Petera, 70 F.3d 1566 (10th Cir. 1995).
6. Alford v. Finch, 155 So.2d 790 (Fla. 1963).
7. Allen v. McClellan, 405 P.2d 405 (N.M. 1965).
8. Sickman v. United States, 184 F.2d 616 (7th Cir. 1950).
9. Jordan v. State, 681 P.2d 346 (Alaska App. 1984).
10. Moerman v. State, 21 Cal. Rptr. 2d 329 (Cal. App. 1993).
11. Parker Land & Cattle Co. v. Wyoming Game & Fish Commission, 845 P.2d 1040 (Wyo. 1993).
12. Cross v. State, 370 P.2d 371 (Wyo. 1962).
13. State v. Vander Houwen, 177 P.3d 93 (Wash. 2008).
14. Cook v. State, 74 P.2d 199 (Wash. 1937).
15. State v. Cleve, 980 P.2d 23 (N.M. 1999).
16. State v. Thompson, 33 P.3d 213 (Idaho App. 2001).
17. Christy v. Hodel, 857 F.3d 1324 (9th Cir. 1988).
18. Conn. Gen. Stat. § 26-47.
19. Dept. of Community Affairs v. Moorman, 664 So.2d 930 (Fla. 1995).
20. State v. Sour Mountain Realty Co., 714 N.Y.S.2d 78 (N.Y. App. Div. 2000).
21. United States *ex rel.* Bergen v. Lawrence, 848 F.2d 1502 (10th Cir. 1988).
22. State Department of Fisheries v. Gillette. 621 P.2d 764 (Wash. App. 1980).
23. An Act for the More Easy Discovery and Conviction of Such as Shall Destroy the Game of This Kingdom, 3 & 4 Wm. & M., ch. 23, § 11 (1692).
24. Justices of the Peace Shall Be Conservators of the Statutes Made Touching Salmons, 17 Rich. 2, ch. 9 (1393).
25. Keeble v. Hickeringill, 11 East 574, 103 English Reports 1127 (Queens Bench 1707).
26. Harrison v. Petroleum Surveys, Inc., 80 So.2d 153 (La. App. 1955).
27. Lenk v. Spezia. 213 P.2d 47 (Cal. App. 1949).
28. Figliuzzi v. Carcajou Shooting Club of Lake KoshKonong, 516 N.W.2d 410 (Wis. 1994).
29. Mikesh v. Peters, 284 N.W.2d 215 (Iowa 1979).

Chapter 5

1. Sir Matthew Hale, "*De Jure Maris*," in Stuart A. Moore, *A History of the Foreshore and the Law Relating Thereto,* 3d ed. (London: Stevens & Haynes, 1888), 374-375.
2. Adams v. Pease, 2 Conn. 481 (Conn. 1818).
3. Yard v. Carman, 3 N.J.L. 2 Pennin. 936 (N.J. 1812).
4. Carson v. Blazer, 2 Binn. 475 (Pa. 1810).
5. Washington Ice. Co. v. Shortall, 101 Ill. 46 (1881).

6. Martin v. Waddell's Lessee, 41 U.S. 367 (1842).
7. Pollard's Lessee v. Hagan. 44 U.S. 212 (1845).
8. Munninghoff v. Wisconsin Conservation Commission, 38 N.W.2d 712 (Wis. 1949).
9. Douglaston Manor, Inc. V. Bahrakis, 678 N.E.2d 201 (N.Y. 1997).
10. Adirondack League Club, Inc. v. Sierra Club, 706 N.E.2d 1192 (N.Y. 1998).
11. Illinois Central R.R. v. Illinois, 146 U.S. 387 (1892).
12. Smith v. Maryland, 59 U.S. (18 How.) 71 (1855).
13. Phillips Petroleum Co. v. Mississippi, 484 U.S. 469 (1988).
14. Idaho v. Coeur d'Alene Tribe, 521 U.S. 261 (1997).
15. Ralph W. Johnson and Russell A. Austin Jr., "Recreational Rights and Titles to Beds on Western Lakes and Streams," 7 *Natural Resources Journal* 1, 24–25 (1967).
16. United States v. W.R. Cress, 243 U.S. 316 (1917).
17. United States v. Rands, 389 U.S. 121 (1967).
18. Lucas v. South Carolina Coastal Council 506 U.S. 1003 (1992).
19. Dardar v. Lafourche Realty Co., 985 F.2d 824 (5th Cir. 1993), *citing* Kaiser Aetna v. United States, 444 U.S. 164 (1979).
20. United States v. Twin City Power Co., 350 U.S. 222 (1956).
21. The Montello, 87 U.S. (20 Wall.) 430 (1874).
22. Atlanta School of Kayaking, Inc. v. Douglasville-Douglas County Water & Sewer Authority, 981 F. Supp. 1469 (N.D. Ga. 1997).
23. Vaughn v. Vermilion Corp., 444 U.S. 206 (1979).
24. Parm v. Shumate, 513 F.3d 135 (5th Cir. 2007).
25. Rivers and Harbors Act, 33 U.S.C. § 401.
26. 1 Stat. 51 (1789).
27. Beacham v. Lake Zurich Property Owners Association, 526 N.E.2d 154 (Ill. 1988).
28. Carnahan v. Moriah Property Owners Association, Inc. 716 N.E.2d 437 (Ind. 1999).
29. People v. Truckee Lumber Co., 48 P. 374 (Cal. 1897).
30. Snyder v. Callaghan, 284 S.E.2d 241 (W.Va. 1981).
31. Springer v. Joseph Schlitz Brewing Co., 510 F.2d 468 (4th Cir. 1975).
32. State v. Haskell, 79 A. 852 (Vt. 1911).
33. 33 U.S.C. §§ 1251-1387.

Chapter 6

1. Lacey Act, 16 U.S.C. §§ 701, 3771-3378.
2. Migratory Bird Treaty Act, 16 U.S.C. § 703-711.
3. Missouri v. Holland, 252 U.S. 416 (1920).

4. Camfield v. United States, 167 U.S. 518 (1897).
5. United States v. Alford, 274 U.S. 264 (1927).
6. Hunt v. United States, 278 U.S. 96 (1928).
7. Kleppe v. New Mexico, 426 U.S. 529 (1976).
8. Wild Free-Roaming Horses and Burros Act, 16 U.S.C. §§ 1331-1340.
9. United States v. Brown, 552 F.2d 817 (8th Cir. 1977).
10. Stupak-Thrall v. United States, 843 F. Supp. 327 (W.D. Mich. 1994).
11. Grand Lake Estates Homeowners Association v. Veneman, 340 F. Supp. 2d 1162 (D. Colo. 2004).
12. Douglas v. Seacoast Products, Inc., 431 U.S. 265 (1977).
13. United States v. Helsley, 615 F.2d 784 (9th Cir. 1979).
14. Endangered Species Act, 16 U.S.C. §§ 1531-1544.
15. Clean Water Act, 33 U.S.C. § 1344.
16. State v. Bontrager, 683 N.E.2d 126 (Ohio App. 1996).
17. District of Columbia v. Heller, 128 S.Ct. 2783 (2008).
18. Takahashi v. Fish & Game Commission, 334 U.S. 410 (1948).
19. Kafka v. Hagener, 176 F. Supp. 2d 1037 (D. Mont. 2001).
20. Schutz v. Thorne, 415 F.3d 1128 (10th Cir. 2005).
21. Alaska Constitutional Legal Defense Conservation Fund, Inc. v. Kempthorne, 198 Fed. Appx. 601 (9th Cir. 2006).
22. Andrus v. Allard, 444 U.S. 51 (1979).
23. Bald and Golden Eagle Protection Act, 16 U.S.C. §§ 668-668d.
24. Migratory Bird Treaty Act, 16 U.S.C. §§ 703-711.
25. Penn Central Transportation Co. v. City of New York, 438 U.S. 104 (1978).
26. Arctic King Fisheries, Inc. v United States, 59 Fed. Cl. 360 (2004).
27. The American Fisheries Act amended the Magnuson-Stevens Fishery Conservation and Management Act, 16 U.S.C. §§ 1801-1882.
28. Wyoming v. United States, 279 F.3d 1214 (10th Cir. 2002).
29. Young v. Coloma-Agaran, 2001 WL 1677259 (D. Hawaii, 2001).
30. Pacific Northwest Venison Producers v. Smitch, 20 F.3d 1008 (9th Cir. 1994).
31. United States v. Earp, 307 F. Supp. 2d 760 (D.S.C. 2003).
32. Geer v. Connecticut, 161 U.S. 519 (1896).
33. Hughes v. Oklahoma, 441 U.S. 322 (1979)
34. Maine v. Taylor, 477 U.S. 131 (1986).
35. McCready v. Virginia, 94 U.S. 391 (1876).
36. Toomer v. Witsell, 334 U.S. 385 (1948).
37. Connecticut v. Crotty, 346 F. 3d 84 (2d Cir. 2003).
38. Baldwin v. Fish & Game Commission, 436 U.S. 371 (1978).
39. Tangier Sound Waterman's Association v. Pruitt, 4 F.3d 264 (4th Cir. 1993).
40. Taulman v. Hayden, 2006 WL 2631914 (D. Kan. 2006).
41. Minnesota *ex rel.* Hatch v. Hoeven, 456 F.3d 826 (8th Cir. 2006).

42. Kafka v. Hagener, 176 F. Supp. 2d 1037 (D. Mont. 2001).
43. Conservation Force, Inc. v. Manning, 301 F.3d 985 (9th Cir. 2002).
44. California Coastal Commission v. Granite Rock Co., 480 U.S. 572 (1987).

Chapter 7

1. The Statute of Westminster II, ch. 47 (1285).
2. An Act for the Better Preservation of the Game, and for Securing Warrens not Inclosed, and the Several Fishings of the Realm, 22 & 23 Car. 2, § 3 (1671).
3. Waltham Black Act, 9 Geo. I, ch. 22 (1723).
4. Humane Society v. New Jersey State Fish & Game Council, 362 A.2d 20 (N.J. 1976).
5. Bean v. McWherter, 953 S.W.2d 197 (Tenn. 1997).
6. State v. Maschell, 677 N.W.2d 551 (S.D. 2004).
7. Wisconsin Citizens Concerned for Cranes and Doves v. Wisconsin Department of Natural Resources, 677 N.W.2d 612 (Wis. 2004).
8. State v. Couch, 147 P.3d 322 (Or. 2006).
9. Armstrong v. State, 958 P.2d 1010 (Wash. App. 1998).
10. Massachusetts Society for the Prevention of Cruelty to Animals v. Division of Fisheries and Wildlife, 651 N.E.2d 388 (Mass. 1995).
11. State v. Bonnewell, 2 P.3d 682 (Ariz. App. 1999).
12. Fund for Animals v Oregon Department of Fish & Wildlife, 765 P.2d 215 (Or. App. 1988).
13. State v. Casano, 95 P.3d 79 (Idaho App. 2004).
14. Arkansas Game & Fish Commission v. Murders, 938 S.W.2d 854 (Ark. 1997).
15. State v. Walsh, 870 P.2d 974 (Wash. 1994).
16. State v. McAffry, 949 P.2d 1137 (Kan. 1997).
17. State v. Morrison, 341 N.W.2d 635 (S.D. 1983).
18. Singleton v. Commonwealth, 740 S.W.2d 159 (Ky. App. 1986).
19. State v. Casano, 95 P.3d 79 (Idaho App. 2004).
20. Corry v. State, 710 So.2d 853 (Miss. 1998).
21. State v. Davis, 448 So.2d 645 (La. 1984).
22. State v. Cloutier, 814 A.2d 966 (Me. 2003).
23. State v. Thompson, 948 P.2d 174 (Idaho App. 1997).
24. State v. Neilson, 660 So.2d 130 (La. App. 1995).
25. State v. Walsh, 870 P.2d 974 (Wash. 1994).
26. State v. Vogt, 55 P.3d 365 (Kan. App. 2002).
27. State v. Bradley, 866 A.2d 242 (N.J. Super.L. 2004).
28. Commonwealth v. Sellinger, 763 A.2d 525 (Pa. Cmnwlth. 2000).
29. State v. Mobbs, 740 A.2d 1288 (Vt. 1999).
30. State v. Bowersmith, 2002 WL 1434057 (Ohio App. 2002).

31. State v. Huebner, 827 P.2d 1260 (Mont. 1992).
32. People v. Gordon, 160 P.3d 284 (Colo. App. 2007).
33. United States v. Hunnicutt, 2006 WL 91765 (W.D. N.C. 2006).
34. State v. Boyer, 42 P.3d 771 (Mont. 2002).
35. State v. McHugh, 630 So.2d 1259 (La. 1994).
36. State v. Romain, 983 P.2d 322 (Mont. 1999).
37. Corry v. State, 710 So.2d 853 (Miss. 1998).
38. Commonwealth v. Russo, 934 A.2d 1199 (Pa. 2007).
39. Rainey v. Hartness, 5 S.W.3d 410 (Ark. 1999).
40. State v. Larsen, 650 N.W.2d 144 (Minn. 2002).
41. G. J. Leasing Co. v. Union Electric Co., 54 F.3d 379, 386 (7th Cir. 1995).
42. Irvine v. Rare Feline Breeding Center, Inc., 685 N.E.2d 120 (Ind. 1997).
43. Oklahoma City v. Hudson, 405 P.2d 178 (Ok. 1965).
44. Pate v. Yeager, 552 S.W.2d 513 (Tex. Civ. App. 1977).
45. Smith v. Jalbert, 221 N.E.2d 744 (Mass. 1966).
46. Briley v. Mitchell, 115 So.2d 851 (La. 1959).
47. Candler v. Smith, 179 S.E. 395 (Ga. App. 1935).
48. Phillips v. Garner, 64 So. 735 (Miss. 1914).
49. Nicholson v. Smith, 986 S.W.2d 54 (Tex. App. 1999).
50. Robison v. Gantt, 673 So.2d 441 (Ala. Civ. App. 1995).
51. Overstreet v. Gibson Product Co., 558 S.W.2d 58 (Tex. Civ. App. 1977).
52. Brunelle v. Signore, 263 Cal. Rptr. 415 (1989).
53. King v. Blue Mountain Forest Association, 123 A.2d 151 (N.H. 1956).
54. Andrews v. Andrews, 88 S.E.2d 88 (N.C. 1955).
55. Carlson v. State, 598 P.2d 969 (1979).
56. Arroyo v. State, 40 Cal. Rptr.2d 627 (Cal. App. 1995).
57. Wamser v. City of St. Petersburg, 339 So.2d 244 (Fla. App. 1976).
58. Claypool v. United States, 98 F. Supp. 702 (S.D. Cal. 1951).
59. Rubenstein v. United States, 338 F. Supp. 654 (N.D. Cal. 1972).
60. Leslie v. State, 502 N.Y.S.2d 825 (App. Div. 1986).
61. 989 P.2d 113 (Wyo. 1999).
62. Sakach v. City of Pittsburgh, 687 A.2d 34 (Pa. Cmnwlth. 1996).
63. Franken v. City of Sioux Center, 272 N.W.2d 422 (Iowa 1978).
64. City and County of Denver v. Kennedy, 476 P.2d 762 (Colo. App. 1970).
65. 83 P.3d 61 (Ariz. App. 2004).

Chapter 8

1. Cherokee Nation v. Georgia, 30 U.S. (5 Pet.) 1 (1831).
2. United States v. Mazurie, 419 U.S. 544, 557 (1975).
3. New Mexico v. Mescalero Apache Tribe, 462 U.S. 324, 331-332 (1983).
4. Mitchell v. United States, 34 U.S. (9 Pet.) 711 (1835).

5. United States v. Williams, 198 U.S. 371 (1905).
6. State v. Shook, 67 P.3d 863 (Mont. 2003).
7. Washington v. Washington Commercial Passenger Fishing Vessel Association, 443 U.S. 658, 690 (1979).
8. United States v. Dion, 476 U.S. 734 (1986).
9. Endangered Species Act, 16 U.S.C. §§ 1531-1544.
10. United States v. Peterson, 121 F. Supp. 2d 1309 (D. Mont. 2000).
11. 163 U.S. 504 (1896).
12. United States v. Williams, 198 U.S. 371 (1905).
13. Minnesota v. Mille Lacs Band of Chippewa Indians, 526 U.S. 172 (1999).
14. Treaty with the Yakama Nation, art. III., June 11, 1855, 12 Stat. 951.
15. United States v. Taylor, 13 P. 333 (Wash. Terr. 1887).
16. United States v. Winans, 198 U.S. 371 (1905).
17. Whitefoot v. United States, 293 F.2d 658 (Ct. Cl. 1961).
18. United States v. Washington, 520 F.2d 676 (9th Cir. 1975).
19. New Mexico v. Mescalero Apache Tribe, 462 U.S. 324 (1983).
20. Nevada v. Hicks, 533 U.S. 353 (2001).
21. New Mexico v. Mescalero Apache Tribe, 462 U.S. 324 (1983).
22. Nevada v. Hicks, 533 U.S. 353 (2001).
23. State v. Utah, 128 P.3d 1211 (2006).
24. State v. Cayenne, 158 P.3d 623 (Wash. App. 2007).
25. Nevada v. Hicks, 533 U.S. 353 (2001).
26. 18 U.S.C. § 1162.
27. State v. Jacobs, 735 N.W.2d 535 (Wis. App. 2007).
28. New Mexico v. Mescalero Apache Tribe, 426 U.S. 324 (1983).
29. 533 U.S. 353 (2001).
30. 315 U.S. 681 (1942).
31. Puyallup Tribe v. Department of Game, 391 U.S. 392 (1968).
32. Department of Game v. Puyallup Tribe, 414 U.S. 44 (1973).
33. Puyallup Tribe, Inc. v. Department of Game, 433 U.S. 165 (1977).
34. Washington v. Washington State Commercial Passenger Fishing Vessel Association, 443 U.S. 658 (1979).
35. United States v. Washington (Phase II), 506 F. Supp. 187 (W.D. Wash. 1980).
36. United States v. State of Washington, 2007 WL 2437166 (W.D. Wash. 2007).
37. Montana v. United States, 450 U.S. 544 (1981).
38. Settler v. Lameer, 507 F.2d 231 (9th Cir. 1974).
39. United States v. Winans, 198 U.S. 371 (1905).

Chapter 9

1. Lacey Act, 16 U.S.C. § 3372(a).
2. United States v. Condict, 2006 WL 1793235 (E.D. Okla. 2006).

3. United States v. Kilpatrick, 347 F. Supp. 2d 693 (D. Neb. 2004).
4. United States v. Gardner, 244 F.3d 784 (10th Cir. 2001).
5. United States v. Carpenter, 933 F.2d 748 (9th Cir. 1991).
6. Migratory Bird Treaty Act, 16 U.S.C. §§ 703-711.
7. United States v. Hale, 113 Fed. Appx. 108 (6th Cir. 2004).
8. United States v. Lewis, 240 F.3d 866 (10th Cir. 2001).
9. United States v. Gay-Lord, 799 F.2d 124 (4th Cir. 1986).
10. United States v. Kern, 2007 WL 4377839 (S.D. Tex. 2007).
11. United States v. Labs of Virginia, 272 F. Supp. 764 (N.D. Ill. 2003).
12. United States v. McNab, 331 F.3d 1228 (11th Cir. 2003).
13. United States v. Lee, 937 F.2d 1388 (9th Cir. 1991).
14. United States v. Anderson, 60 Fed. Appx. 761 (10th Cir. 2003).
15. 16 U.S.C. § 3373(a)(1) (civil penalties); *id.* § 3373 (b)(1) (criminal penalties).
16. United States v. Kraft, 2005 WL 57813 (D. Minn. 2005).
17. United States v. Senchenko, 133 F.3d 1153 (9th Cir. 1998).
18. United States v. LeVeque, 283 F.3d 1098 (9th Cir. 2002).
19. United States v. Miranda, 835 F.2d 830 (11th Cir. 1988).
20. United States v. Santillan, 243 F.3d 1125 (9th Cir. 2001).
21. United States v. Hale, 113 Fed. Appx. 108 (6th Cir. 2004).
22. United States v. 144,774 Pounds of Blue King Crab, 410 F.3d 1131 (10th Cir. 2005).
23. 16 U.S.C. §§ 3371(g), 3372(a)(2)(C).
24. Convention for the Protection of Migratory Birds, Aug. 16, 1916, United States–Great Britain, 39 Stat. 1702, T.S. No. 628.
25. Convention for the Protection of Migratory Birds and Game Mammals, Feb. 7, 1936, United States–Mexico, 50 Stat. 1311, T.S. No. 912; Convention for the Protection of Migratory Birds and Birds in Danger of Extinction, and Their Environment, Mar. 4, 1972, United States–Japan, 25 U.S.T. 3329; Convention Concerning the Conservation of Migratory Birds and Their Environment, Nov. 18, 1976, United States–U.S.S.R., 29 U.S.T. 4647, T.I.A.S. No. 9073.
26. Migratory Bird Treaty Act, 16 U.S.C. §§ 703-711.
27. Migratory Bird Treaty Act, 16 U.S.C. § 703.
28. United States v. Darst, 726 F. Supp. 286 (D. Kan. 1989).
29. Migratory Bird Treaty Act, 16 U.S.C. § 711.
30. Noe v. Henderson, 456 F.3d 868 (8th Cir. 2006).
31. Koop v. United States, 296 F.2d 53 (8th Cir. 1961).
32. 50 C.F.R. § 21.13.
33. Hill v. Norton, 275 F.3d 98 (D.C. Cir. 2001).
34. Migratory Bird Treaty Act, 16 U.S.C. § 703(b).
35. The Fund for Animals v. Norton, 374 F. Supp. 2d 91 (D.D.C. 2005).
36. 70 Fed. Reg. 12710 (Mar. 15, 2005).

37. United States v. Winddancer, 435 F. Supp. 2d 687 (M.D. Tenn. 2006).
38. United States v. Morgan, 311 F.3d 611 (5th Cir. 2002).
39. United States v. FMC Corp., 572 F.2d 902 (2nd Cir. 1978).
40. United States v. Corbin Farm Service, 444 F. Supp. 510 (E.D. Cal. 1978).
41. United States v. Rollins, 706 F. Supp. 742 (D. Idaho 1989).
42. United States v. WCI Steel, Inc., 2006 WL 2334719 (N.D. Ohio 2006).
43. City of Sausalito v. O'Neill, 386 F.3d 1186 (9th Cir. 2004).
44. Seattle Audubon Society v. Evans, 952 F.2d 297 (9th Cir. 1991).
45. 16 U.S.C. § 704(b).
46. United States v. Strassweg, 143 Fed. Appx. 665 (6th Cir. 2005).
47. 50 C.F.R. § 20.11.
48. Falk v. United States, 452 F.3d 951 (8th Cir. 2006).
49. Center for Biological Diversity v. Pirie, 191 F. Supp. 2d 161 (D.D.C. 2002).
50. Migratory Bird Treaty Act, 16 U.S.C., § 703.
51. United States v. Winnett, 2003 WL 21488645 (D. Mass. 2003).
52. Migratory Bird Treaty Act, 16 U.S.C. § 707(b).
53. United States v. Zak, 486 F. Supp.2d 208 (D. Mass. 2007).
54. Endangered Species Act, 16 U.S.C. § 1540.
55. United States v. Zak, 486 F. Supp. 2d 208 (D. Mass. 2007).
56. Migratory Bird Treaty Act, 16 U.S.C. § 668.
57. United States v. Allard, 397 F. Supp. 429 (D. Mont. 1975).
58. United States v. Zak, 486, F. Supp. 2d 208 (D. Mass. 207).
59. United States v. Moon Lake Electric Association, Inc., 45 F. Supp. 2d 1070 (D. Colo. 1999).

Chapter 10

1. Kleppe v. New Mexico, 426 U.S. 529 (1976).
2. National Environmental Policy Act, 42 U.S.C. §§ 4321, 4331-4335.
3. Wilderness Act, 16 U.S.C., §§ 1131-1136.
4. National Wildlife Refuge Administration Act, 16 U.S.C. §§ 668dd-668ee.
5. Migratory Bird Conservation Act, 16 U.S.C. § 715d.
6. Refuge Recreation Act, 16 U.S.C. §§ 460k to 460k-4.
7. Act of Aug. 12, 1949, ch. 421, 63 Stat. 599.
8. Act of Aug. 1, 1958, Pub. L. 85-585, 72 Stat. 486.
9. 16 U.S.C. §§ 668dd-668ee.
10. National Wildlife Refuge System Improvement Act, Pub. L. 105-57, 111 Stat. 1254, codified into the National Wildlife Refuge Administration Act at 16 U.S.C. §§ 668dd-668ee.
11. 16 U.S.C. § 668dd(a)(2).
12. 66 Fed. Reg. 3810 (Jan. 16, 2001).

13. Provisions specific to individual refuges are found in various provisions of the Code of Federal Regulations, available online.
14. Schwenke v. Secretary of the Interior, 720 F.2d 571 (9th Cir. 1983).
15. 16 U.S.C. §§ 668dd-668ee.
16. 16 U.S.C. § 668dd(e).
17. Moore v. Kempthorne, 464 F. Supp. 2d 519 (E.D. Va. 2006).
18. Nibrara River Ranch, L.L.C. v. Huber, 373 F.3d 881 (8th Cir. 2004).
19. Defenders of Wildlife v. Andrus, 455 F. Supp. 446 (D.D.C. 1978).
20. Stevens County v. U.S. Department of Interior, 507 F. Supp. 2d 1127 (E.D. Wash. 2007).
21. 16 U.S.C. § 668dd(m).
22. 16 U.S.C. §§ 668dd(a)(4)(E), (m).
23. Wyoming v. United States, 279 F.3d 1214 (10th Cir. 2002).
24. National Audubon Society v. Davis, 307 F.3d 835 (9th Cir. 2002).
25. 18 U.S.C. § 41.
26. 16 U.S.C. § 668dd(f).
27. 16 U.S.C. § 668dd(c).
28. 16 U.S.C. § 668dd(e).
29. Wyoming v. Livingston, 443 F.3d 1211 (10th Cir. 2006).
30. This issue is explored in detail in Eric T. Freyfogle, "The Wildlife Refuge and the Land Community," 44 *Natural Resources Journal,* 1027–40 (2004).
31. 16 U.S.C. § 668dd(a)(2).
32. 16 U.S.C. § 668dd(a)(4)(B).
33. 16 U.S.C. § 668dd(a)(4)(C).
34. 66 Fed. Reg. 3810-3818 (Jan. 16, 2001).
35. Id.
36. Id.
37. 66 Fed. Reg. 3810, 3820 (Jan. 16, 2001).
38. 16 U.S.C. § 1.
39. 36 C.F.R. § 2.2.
40. Alaska Wildlife Alliance v. Jensen, 108 F.3d 1065 (9th Cir. 1997).
41. National Rifle Association v. Potter, 628 F. Supp. 903 (D.D.C. 1986).
42. Michigan United Conservation Clubs v. Lujan, 949 F.2d 202 (6th Cir. 1991).
43. Fund for Animals v. Mainella, 294 F. Supp. 2d 46 (D.D.C. 2003).
44. Organic Act, 16 U.S.C. §§ 1-3.
45. National Wildlife Federation v. National Park Service, 669 F. Supp. 384 (D. Wyo. 1987).
46. Conservation Law Foundation v. Clark, 590 F. Supp. 1467 (D. Mass. 1984), affd. 864 F.2d 954 (1st Cir. 1989).
47. Sierra Club v. Babbitt, 69 F. Supp. 2d 1202 (E.D. Cal. 1999).

48. Southern Utah Wilderness Alliance v. National Park Service, 387 F. Supp. 2d 1178 (D. Utah 2005).
49. Southern Utah Wilderness Alliance v. Dabney, 7 F. Supp. 2d 1205 (D. Utah 1998).
50. National Forest Management Act, 16 U.S.C. §§ 1600-1616.
51. Federal Land Policy and Management Act, 43 U.S.C. §§ 1701-1784.
52. 16 U.S.C. § 475.
53. 16 U.S.C. §§ 528-531.
54. 16 U.S.C. § 530.
55. 16 U.S.C. § 528.
56. 16 U.S.C. §§ 1600-1616.
57. 16 U.S.C. § 1604(g)(3)(B).
58. Now repealed (in 2005), the provisions appeared at 36 C.F.R. §§ 219.12, 219.19, and 291.26.
59. Sierra Club v. Marita, 46 F.3d 606 (7th Cir. 1995).
60. 70 Fed. Reg. 1023-01 (Jan. 5, 2005).
61. Id.
62. Ohio Forestry Association v. Sierra Club, 523 U.S. 726 (1998).
63. Norton v. Southern Utah Wilderness Alliance, 542 U.S. 55 (2004).
64. 36 C.F.R. § 219.10.
65. 36 C.F.R. § 219.10(b).
66. 36 C.F.R. § 219.10(b)(2).
67. Citizens for Better Forestry v. U.S. Dept. of Agriculture, 481 F. Supp. 2d 1059 (N.D. Cal. 2007).
68. Citizens for Better Forestry v. U.S. Dept. of Agriculture, 481 F. Supp. 2d 1059 (N.D. Cal. 2007).
69. 16 U.S.C. § 1604(g)(3)(F).
70. Codifed as amended at 43 U.S.C. §§ 315-315r.
71. 43 U.S.C. §§ 1701-1784.
72. 16 U.S.C. §§ 1331-1340.
73. 16 U.S.C. § 670g.
74. Federal Land Policy and Management Act, 43 U.S.C. § 1732(b).
75. Alaska v. Andrus, 429 F. Supp. 958 (D. Alaska1977).
76. Alaska v. Andrus, 591 F.2d 537 (9th Cir. 1979); Defenders of Wildlife v. Andrus, 627 F.2d 1238 (D.C. Cir. 1980).

Chapter 11

1. 16 U.S.C. §§ 1531-1544.
2. 16 U.S.C. § 1531(b).

3. Endangered Species Act, 16 U.S.C., §§ 1538(a)(1)(B)-(C), 1532(19).
4. 16 U.S.C. § 1536(a)(2).
5. 16 U.S.C. § 1532(f).
6. 16. U.S.C. § 1533(f).
7. 16 U.S.C. § 1532(16).
8. Alabama-Tombigbee Rivers Coalition v. Kempthorne, 477 F.3d 1250 (11th Cir. 2007).
9. Institute for Wildlife Protection v. Norton, 174 Fed. Appx. 363 (9th Cir. 2006).
10. 61 Fed. Reg. 4722 (Feb. 7, 1996).
11. Northwest Ecosystem Alliance v. United States Fish and Wildlife Service, 475 F.3d 1136 (9th Cir. 2007).
12. National Wildlife Federation v. Norton, 386 F. Supp. 2d 553 (D. Vt. 2005); Defenders of Wildlife v. Norton, 354 F. Supp. 2d 1156 (D. Or. 2006).
13. 16 U.S.C. § 1532(6).
14. 16 U.S.C. § 1532(20).
15. 16 U.S.C. § 1533(b).
16. 16 U.S.C. § 1533.
17. 16 U.S.C. §1533(a)(1).
18. 16 U.S.C. §1533(b)(1)(A).
19. 16 U.S.C. § 1533(b)(1)(A).
20. Western Watersheds Project v. Foss, 2005 WL 2002473 (D. Idaho 2005).
21. Defenders of Wildlife v. Norton, 258 F.3d 1136 (9th Cir. 2001).
22. Center for Biological Diversity v. Norton, 411 F. Supp. 2d 1271 (D. N.M. 2005).
23. Solicitor's Opinion M-37013 (Mar. 16, 2001).
24. 16 U.S.C. § 1533(a)(1).
25. Policy for the Evaluation of Conservation Efforts When Making Listing Decisions, 68 Fed. Reg. 15100-02 (Mar. 28, 2003).
26. 16 U.S.C. § 1533(a)(1)(E).
27. Center for Biological Diversity v. Norton, 254 F.3d 833 (9th Cir. 2001).
28. 68 Fed. Reg. 15113.
29. Western Watersheds Project v. Foss, 2005 WL 2002473 (D. Idaho 2005).
30. 16 U.S.C. § 1533(b)(3)(A).
31. 16 U.S.C. § 1533(b)(3)(A).
32. 16 U.S.C. § 1533(b)(3)(B).
33. 16 U.S.C. § 1533(b)(6).
34. 16 U.S.C. § 1533(b)(3)(B)(iii).
35. 16 U.S.C. § 1533(b)(3)(C)(iii).
36. Center for Biological Diversity v. Kempthorne, 466 F.3d 1098 (9th Cir. 2006).
37. California Native Plant Society v. Norton, 2005 WL 768444 (D.D.C. 2005).
38. 16 U.S.C. § 1533(a)(3).
39. 16 U.S.C. § 1532(5)(A).

40. 16 U.S.C. § 1534(b)(2).
41. 16 U.S.C. § 1533(b)(6)(C).
42. 16 U.S.C. § 1536(a)(2).
43. Sierra Club v. U.S. Fish & Wildlife Service, 245 F.3d 434 (2001).
44. New Mexico Cattle Growers Association v. U.S. Fish & Wildlife Service, 248 F.3d 1277 (10th Cir. 2001).
45. Arizona Cattle Growers' Association v. Kempthorne, 534 F. Supp. 2d 1013 (D. Ariz. 2008).
46. Home Builders Association of Northern California v. United States Fish and Wildlife Service, 2007 WL 201248 (E.D. Cal. 2007).
47. 16 U.S.C. § 1533(b)(3)(D).
48. Schoeffler v. Kempthorne, 493 F. Supp. 2d 805 (W.D. La. 2007).
49. Alabama-Tombigbee Rivers Coalition v. Kempthorne, 477 F.3d 1250 (11th Cir. 2007).

Chapter 12

1. Endangered Species Act, 16 U.S.C. § 1536(a)(2).
2. 16 U.S.C. § 1536(a)(1).
3. 16 U.S.C. § 1536(a)(2).
4. 437 U.S. 153 (1978).
5. 50 C.F.R. § 402.02.
6. 524 F.3d 917 (9th Cir. 2008).
7. 509 F.3d 1310 (10th Cir. 2007).
8. Oregon Natural Resources Council v. Hallock, 2006 WL 3463432 (D. Or. 2006).
9. 127 S.Ct. 2518 (2007).
10. 50 C.F.R. § 402.12(b).
11. 50 C.F.R. § 402.12 (a), (f).
12. National Wildlife Federation v. National Marine Fisheries Service, 422 F.3d 782 (9th Cir. 2005); National Wildlife Federation v. Norton, 332 F. Supp. 2d 170 (D.D.C. 2004); Heartwood v. Kempthorne, 2007 WL 1795296 (S.D. Ohio 2007).
13. National Wildlife Federation v. Norton, 332 F. Supp. 2d 170 (D.D.C. 2004).
14. New Mexico Cattle Growers Association v. U.S. Fish & Wildlife service, 248 F.3d 1277 (10th Cir. 2001); Defenders of Wildlife v. Babbitt, 130 F. Supp. 2d 121 (D.D.C. 2001).
15. 50 C.F.R. § 402.14(g).
16. Conner v. Burford, 848 F.2d 1441 (9th Cir. 1988).
17. National Wildlife Federation v. Norton, 332 F. Supp. 2d 170 (D.D.C. 2004).
18. Florida Key Deer v. Brown, 364 F. Supp. 2d 1345 (S.D. Fla. 2005).
19. Pacific Coast Federation of Fishermen's Associations v. United States Bureau of Reclamation, 426 F.3d 1082 (9th Cir. 2005).

20. Natural Resources Defense Council v. Kempthorne, 506 F. Supp. 2d 322 (E.D. Cal. 2007).
21. 485 F. Supp. 2d 1190 (D. Or. 2007).
22. 50 C.F.R. § 402.16.
23. Forest Guardians v. Johanns, 450 F.3d 455 (9th Cir. 2006); Southwest Center for Biological Diversity v. Bartel, 470 F. Supp. 2d 1118 (S.D. Cal. 2006).
24. Aluminum Company of America v. Administrator, Bonneville Power Administration, 175 F.3d 1156 (9th Cir. 1999).
25. Resources Ltd. v. Robertson, 35 F.3d 1300 (9th Cir. 1993).
26. Aluminum Company of America v. Administrator, Bonneville Power Administration, 175 F.3d 1156 (9th Cir. 1999).
27. 16 U.S.C. § 1538(a)(1)(B).
28. 16 U.S.C. § 1532(19).
29. 50 C.F.R. § 17.3.
30. 515 U.S. 687 (1995).
31. 16 U.S.C. § 1533(d).
32. 755 F.2d 608 (8th Cir. 1985).
33. 16 U.S.C. § 1536(b)(4).
34. Oregon Natural Resources Council v. Allen, 476 F.3d 1031 (9th Cir. 2007).
35. 16 U.S.C. § 1539(a)(1)(A).
36. 16 U.S.C. § 1539(a)(1)(B).
37. 16 U.S.C. § 1539(a)(2)(B).
38. Spirit of Sage Council v. Kempthorne, 511 F. Supp. 2d 31 (D.D.C. 2007).
39. 16. U.S.C. § 1536(a)(1).
40. Id.
41. House v. United States Forest Service, 974 F. Supp. 1022 (E.D. Ky. 1997).
42. Defenders of Wildlife v. Flowers, 2003 WL 22143271 (D. Ariz. 2003).
43. Oregon Natural Resources Council Fund v. U.S. Army Corps of Engineers, 2003 WL 22143271 (D. Or. 2003).
44. 16 U.S.C. § 1533(f).
45. Sierra Club v. Lujuan, 36 Envt'l Rep. Case. 1533 (W.D. Tex. 1993).
46. Grand Canyon Trust v. Norton, 2006 WL 167560 (D. Ariz. 2006).
47. Conservation Northwest v. Kempthorne, 2007 WL 1847143 (W.D. Wash. 2007).
48. 16 U.S.C. § 1536(e).
49. 16 U.S.C. § 1539 (j).
50. Wyoming Farm Bureau Federation v. Babbitt, 199 F.3d 1224 (2000).
51. 16 U.S.C. § 1538(a)(1).
52. 16 U.S.C. § 1539(e).
53. 16 U.S.C. § 1540.

Chapter 13

1. *Millennium Ecosystem Assessment, Ecosystems and Human Well-Being: Our Human Planet* (Washington, DC: Island Press, 2005).
2. "Planning for Wildlife," in Aldo Leopold, *For the Health of the Land,* ed. J. B. Callicott and E. Freyfogle (Washington, DC: Island Press, 1999).
3. Tennessee Environmental Council, Inc. v. Bright Par 3 Associates, 2004 WL 419720 (Tenn. App. 2004).
4. State v. Sour Mountain Realty, Inc., 714 N.Y.S.2d 78 (N.Y. App. Div., 2 Dept. 2000).
5. Southview Associates, Ltd v. Bongartz, 980 F.2d 84 (2d Cir. 1992).
6. Gardner v. New Jersey Pinelands Commission, 593 A.2d 251 (N.J. 1991).
7. Holyoke Co. v. Lyman, 82 U.S. (15 Wall.) 500 (1872); West Point Water Power & Land Improvement Co. v. Moodie, 66 N.W. 6 (Neb. 1896); People v. Parker, 111 Ill. 581 (1884).

Further Reading

Baur, Donald C., and William Robert Irvin. *Endangered Species Act: Law, Policy, and Perspectives.* Chicago: American Bar Association, 2002. (A second, revised edition is in the publication process.)

Bean, Michael J., and Melanie J. Rowland. *The Evolution of National Wildlife Law,* 3d ed. Westport, CT: Praeger, 1997.

Czech, Brian, and Paul R. Krausman. *The Endangered Species Act: History, Conservation Biology, and Public Policy.* Baltimore: Johns Hopkins Press, 2001.

Dunlap, Thomas R. *Saving America's Wildlife: Ecology and the American Mind, 1850–1990.* Princeton, NJ: Princeton University Press, 1991.

Fischman, Robert L. *The National Wildlife Refuges: Coordinating a Conservation System Through Law.* Washington, DC: Island Press, 2003.

Goble, Dale D., J. Michael Scott, and Frank W. Davis, eds. *The Endangered Species Act at Thirty.* Vol. 1, *Renewing the Conservation Promise.* Washington, DC: Island Press, 2006.

Lund, Thomas A. *American Wildlife Law.* Berkeley and Los Angeles: University of California Press, 1980.

McKinstry, Robert B. *Biodiversity Conservation Handbook: State, Local and Private Protection of Biological Diversity.* Washington, DC: Environmental Law Institute, 2006.

Murchison, Kenneth M. *The Snail Darter Case: TVA versus the Endangered Species Act.* Lawrence: University Press of Kansas, 2007.

Musgrave, Ruth. *State Wildlife Laws Handbook.* Washington, DC: Government Institutes, 1993.

Nagle, John Copeland, and J. B. Ruhl. *The Law of Biodiversity and Ecosystem Management,* 2d ed. New York: Foundation Press, 2006.

Ruhl, J. B., Steven Kraft, and Christopher L. Lant. *The Law and Policy of Ecosystem Services.* Washington, DC: Island Press, 2007.

Scott, J. Michael, Dale D. Goble, and Frank W. Davis, eds. *The Endangered Species Act at Thirty.* Vol. 2, *Conserving Biodiversity in Human-Dominated Landscapes.* Washington, DC: Island Press, 2006.

Sellars, Richard West. *Preserving Nature in the National Parks: A History.* New Haven: Yale University Press, 1997.

Sigler, William F. *Wildlife Law Enforcement,* 4th ed. New York: McGraw-Hill Science, 1994.

Stanford Environmental Law Society. *The Endangered Species Act.* Stanford, CA: Stanford University Press, 2001.

Tober, James A. *Who Owns Wildlife? A Political Economy of Conservation in Nineteenth-Century America.* Westport, CT: Greenwood Press, 1981.

Trefethen, James B. *An American Crusade for Wildlife.* New York: Winchester Press, 1975.

Index

About the Authors

Eric T. Freyfogle is Max L. Rowe Professor of Law at the University of Illinois College of Law, where he has taught for twenty-five years in the areas of natural resources, property and land use law, environmental law and policy, wildlife law, and conservation thought. His various writings include *The Land We Share: Private Property and the Common Good; Bounded People, Boundless Lands: Envisioning a New Land Ethic; Natural Resources Law: Private Rights and College Governance,* and, with Dale D. Goble, *Wildlife Law: Cases and Materials.* He has long been active in state and local conservation efforts, including several terms as president of the Illinois affiliate of the National Wildlife Federation.

Dale D. Goble is the Margaret Wilson Schimke Distinguished Professor of Law at the University of Idaho, where his teaching and research focus on the intersection of natural resource law and policy, constitutional law, and history. He has written numerous articles and essays. His book projects include, with Eric T. Freyfogle, *Wildife Law: Cases and Materials;* and with J. Michael Scott and Frank W. Doris, two volumes on the ESA: *The Endangered Species Act at Thirty: Renewing the Conservation Promise* and *The Endangered Species Act at Thirty: Conserving Biodiversity in Human-Dominated Landscapes.*